CW00501994

# Tomorrow's People

ALSO BY PAUL MORLAND

*Demographic Engineering: Population Strategies in Ethnic Conflict*

*The Human Tide: How Population Shaped the Modern World*

# Tomorrow's People

*The Future of Humanity in Ten Numbers*

PAUL MORLAND

PICADOR

First published 2022 by Picador
an imprint of Pan Macmillan
The Smithson, 6 Briset Street, London EC1M 5NR
*EU representative:* Macmillan Publishers Ireland Limited, 1st Floor,
The Liffey Trust Centre, 117–126 Sheriff Street Upper,
Dublin 1 D01 YC43
Associated companies throughout the world
www.panmacmillan.com

ISBN 978-1-5290-4599-4



1 3 5 7 9 8 6 4 2

A CIP catalogue record for this book is available from the British Library.

Typeset in Garamond Pro by Jouve (UK), Milton Keynes
Printed and bound by CPI Group (UK) Ltd, Croydon, CR0 4YY

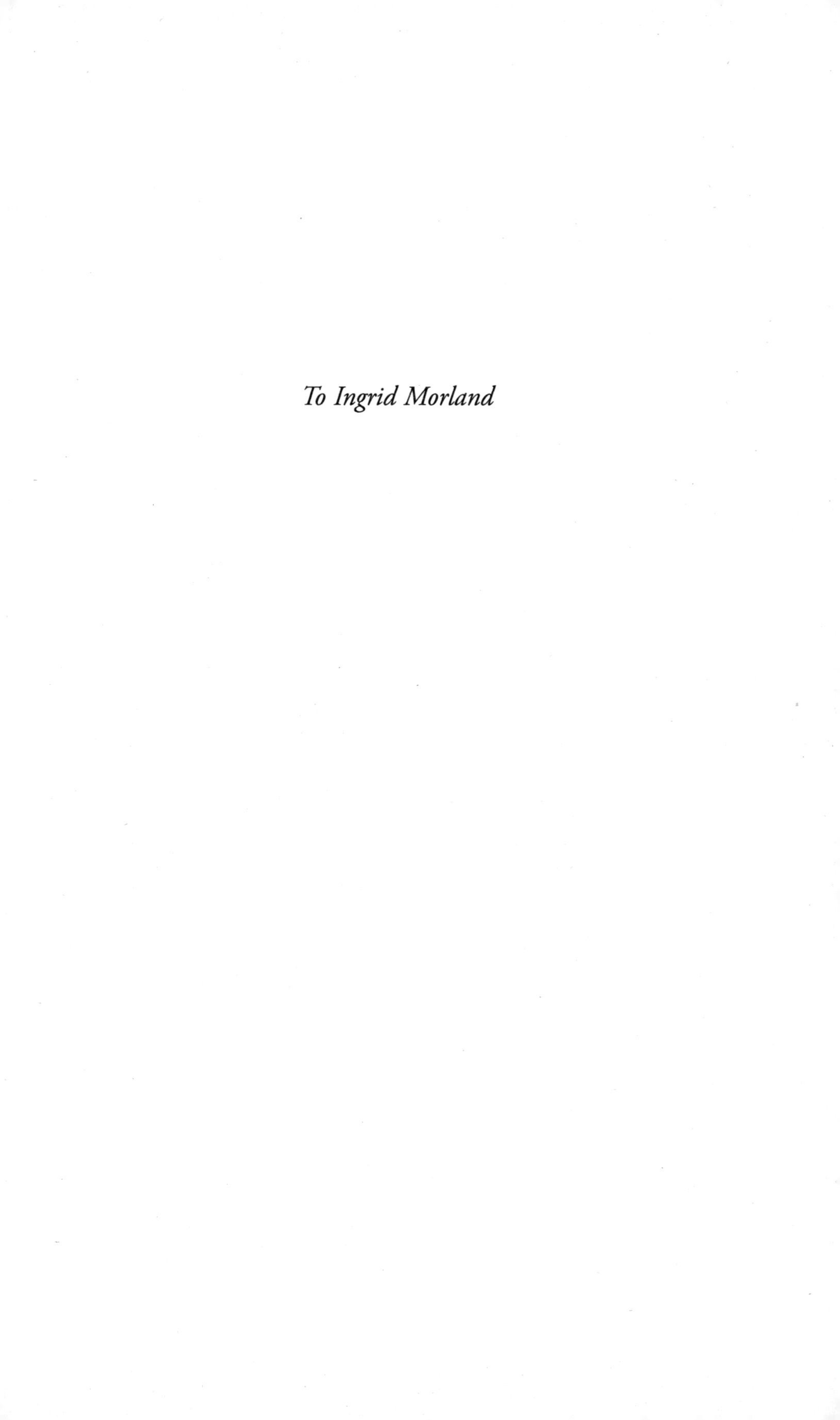

*To Ingrid Morland*

*Contents*

# Introduction

## *Today's People*

The great historical population trends that made us who we are today continue to work on us, shaping our present and future as much as they have shaped our past. The European colonization and domination of the world, which seemed so unshakeable at the end of the nineteenth century, could not have arisen without the continent's population explosion and the exodus of people that it fuelled. Neither the United States nor the Soviet Union would have become twentieth-century superpowers had their populations not ballooned beyond those of their European rivals. Equally, China would not have become America's competitor for global predominance without a population in the hundreds of millions. And nor would India be seen as a coming power were its population not significantly more than a billion.

Just as history's expansions have demographic roots, the same is true of its great reversals. Russia's loss of pre-eminence within the Soviet Union, and the fall of the Soviet Union itself, were in no small part the result of population change. Japan would not now be seen as the 'land of the setting sun' had its population in the 1990s been young, vigorous and growing, as it was when it ascended to world-power status a hundred years earlier. Instead, by the end of the twentieth century, it was a greying country with a declining population and a stagnant economy. And much of the Middle East, from Iraq to Yemen and Libya, would not be in political turmoil if the region were not full of young people with no prospect of economic betterment. None of the great events that dominate the headlines – mass migration, stagnant economies or populism, from the result of the Brexit referendum to Donald Trump's election in the US to Viktor Orbán's Hungary – can be understood without grasping the major population changes that underlie them.[1]

Just as population has moulded our past, it is also shaping our future. Demography may not be destiny, but it is powerful and fast-changing. Europe, which once saw great outward population flow, is now experiencing mass immigration. Where populations were once young, they are now growing old. Countries like Italy which were once renowned for the prodigious sizes of their families now have far fewer children. Countries where a third of babies would once not have lived to see their first birthdays now have infant mortality rates of barely two cases in every thousand. Where once people had little or no formal learning, illiteracy has been consigned to the margins, and where once people went hungry, they have grown obese. Today's people are profoundly different from yesterday's, and tomorrow's people will be different again.

To most of us, the influence of demography on our future is far from obvious. It makes more sense if we divide demographic history into three phases – pre-modern, modern and post-modern – and understand that the process is similar everywhere, though the starting point and rate of change differ from place to place. Communities, countries and even continents are at different points on the journey and are progressing at different speeds, but they are all on the same path.

## *Pre-Modern*

For the vast majority of history, we have been at the mercy of nature in matters of life and death. The sexual impulse determined that men and women would engage in intercourse; although reliable contraception did not exist, there have since ancient times been attempts to decouple sex from pregnancy, some of which were more effective than others. In some places, infanticide was common. Unwanted children were abandoned, or thinned in number through a testing process as in ancient Sparta. Prolonged breastfeeding of the youngest child increased the chance of delaying the next one, and the timing of sex within the menstrual cycle also had some effect. In some cultures, such as Catholic Europe in the Middle Ages, large numbers of people were removed from the reproductive pool, at least in principle, through the institution of priestly celibacy, monasteries and nunneries.

However, the world's population rose steadily over the long term, perhaps quadrupling in the eighteen or so centuries that separated Julius Caesar and Queen Victoria.[2] But high death rates offset high birth rates and kept overall population numbers from what otherwise would have been much faster growth. A civilization with improving technology that was enjoying relative peace might see its population grow, only for it to fall back again. Medieval Europe is the classic example: there was population expansion as land was drained and new ploughing techniques were adopted, but crop failure in the 1310s and the Black Death in the 1340s pushed it down again.[3] China experienced similar golden years of population growth followed by setbacks.

Transport in pre-modern times was rudimentary and expensive. Moving food around to keep large numbers of people alive was often not economic, particularly where overland haulage was required, while duties were often payable, making it even less practical.[4] So people mostly depended on supplies from the immediate vicinity. A poor harvest meant hunger, and a complete crop failure could mean starvation or migration in search of food. And if famine and disease did not keep the population in check, wars and massacres could decimate numbers; seventeenth-century Germany lost around a third of its population during the Thirty Years War, while more than a tenth of the Chinese population died during the fall of the Ming Dynasty.[5]

Births were either unchecked or repressed in crude, unreliable ways, or sometimes by social practices like late marriage, while death travelled freely through populations, especially among the very young. Someone reaching their eightieth birthday stood a better chance of making it to their next birthday than a newborn baby stood of surviving to the age of one. This, roughly, was the state of humanity in pre-modern demography.

### *Modern*

European historians often choose the late fifteenth century as the dividing line between the medieval and the early modern period.[6] Moveable type and printing, originally a Chinese invention, began radically to reduce the cost of learning and created a literate class that benefited from the rapid circulation of ideas. Europeans trying to reach Asia stumbled on the Americas, opening up whole new vistas to explore while spelling demographic doom for their inhabitants. Islam was expelled from Spain but gained a foothold in the Balkans, particularly after the fall of Constantinople to the Ottomans. And the unity of Western Christendom was about to be shattered by the Reformation.

However, none of these changes fundamentally affected the pre-modern population regime. The circulation of precious metal determined who had buying power, while new crops, particularly the potato, gradually transformed life in particular parts of Europe by providing access to cheap carbohydrates. But the key change in population began three hundred years later in one small corner of Europe.

In the late eighteenth century Thomas Malthus responded to optimistic Enlightenment thinkers by providing a description of the population system as it existed up to his day. His 1798 book *An Essay on the Principle of Population* suggested that population, which unconstrained would grow exponentially, was bound to be held in check by a much slower growth in the output of food. But just as Malthus was setting out his system as a God-designed constant, it began to change.[7] Improvements in food supply, public health and medicine were patchy and rudimentary by today's standards, but they were substantial enough to bring down the death rate. The birth rate, meanwhile, remained high and even rose for a time; the resultant accumulation of people was large enough to drive huge population growth at home while also supplying settlers for the United States, Canada and Australasia. This began what is now called the 'demographic transition'.[8]

Following this British head start, other countries soon followed, within Europe at first and then beyond. As the first phase of the transition caught on elsewhere, the next phase began in Britain. Birth rates flagged, bringing the era of population expansion to an end. In the twentieth century, a more educated population that experienced fewer child deaths and had access to improving and affordable contraception was able to choose to have smaller families; during the inter-war period, families with two children became the norm in much of Europe and North America. The demographic transition appeared to be complete; it had been a shift from high fertility and high mortality in a small population to low fertility and low mortality in a large population. Nobody expected the post-war baby boom in North America and parts of Europe, but by the late 1960s it was coming to an end, with fertility rates falling back to, and then below, the 'replacement' level of just over two children per woman.

Just as the first phases of the transition began in Europe and North America before going global, so it was with subsequent phases. Japan was the first country with a non-European population to move from industrialization and urbanization to falling mortality rates, population growth and falling birth rates at the end of the nineteenth century, after which the demographic transition became a truly global phenomenon.

This process has been condemned as a Eurocentric attempt to impose a preference for small families, the medicalization of childbirth and the technological resistance to death on the rest of the world. But if the systems of modernity were imposed by the West, their recipients have embraced them warmly. I'm glad that my wife and I had choices about how many children we had and that we can expect to live into our eighties or beyond, and I'm glad that others can enjoy similar choice and longevity. But even if I lamented the loss of the pre-modern demographic regime, the world would continue getting on with it, regardless of my preferences.

Some countries have barely started on the journey of demographic modernization. Only in recent decades have mortality rates plummeted in much of Africa, and fertility rates in many African nations are still around six children per woman, which is close to what we would have found in pre-modernity. While pre-modern demography was a *state*, a condition of vast breeding and fast dying, demographic modernity is a *process*, a journey towards smaller families and longer lives. It is closely linked with economic, technical and educational progress, the rise of industry, the development of transport and the spread of literacy and education. Much of the world is still going through these stages, but where that process is over, the question is: what will come next?

### *Post-Modern*

Most of the world has either made it through the demographic transition or is well on its way to doing so. This transition was once closely linked to economic progress. As people became richer, better educated and more urbanized, birth rates and death rates declined.[9] Now, instead of accompanying industrial development and economic progress, demographic modernity has leapt ahead of it; families are small and life expectancy is long even in countries that are poor. Sri Lankans live almost as long as Americans, on a fraction of the income. People in Mauritius have on average half a child *fewer* than those in Ireland, although they earn much less. By the start of the twenty-first century, Moroccan women were having well below three children each, even though the majority of them were still illiterate.[10] The decoupling of population and economics provides a hint of what is to come.

The end of the demographic transition does not mean that demographic history is finished; we are witnessing the emergence of *post-modern* demography. Now that modern conditions have been in place long enough to be taken for granted, some people are choosing to have larger families. This is not about economics, industrialization, urbanization or access to contraception, but rather a reflection of culture, values and religion. In some places, demography – and specifically fertility – is driven by ideals rather than material conditions. As demographic forces drive change in the world, we're seeing a shift away from economics to ideas and ideals. It was Marx who pointed out that material conditions drove history, but demography is turning his theory on its head. As populations everywhere are enjoying long life and low mortality, it is fertility that differentiates one community or nation from another, and this is increasingly the product of hopes and fears, aspirations and values rather than material conditions.[11] Amish women in Ohio, for example, have around three times as many children as the average woman in the state, not because of what their families earn but because of what they believe.

Demographic modernization has a parallel in economics; the slowdown in growth among the most advanced economies is regarded by some as the inevitable conclusion of a process that could not continue forever.[12] We find our post-modern demography being reflected in our politics. Identity and age are more important than class – the values you hold play a bigger role in determining how you vote than how you earn your living. Your age is more likely to shape your outlook than your position in the economic pecking order.

There has been talk of a second demographic transition.[13] The theory is that fertility rates will inevitably fall below replacement levels – and permanently – as people prioritize individual pursuits above having a family.[14] Traditional lifestyles will break down and alternatives will proliferate, as people cool on the prospect of marriage and childbearing. As a result, populations will grow smaller and get older, with immigration making up shortfalls in the workforce and societies undergoing a profound ethnic shift. Like the first one, this so-called second demographic transition will initially be Western but will then go global.[15]

However, scratch beneath the surface and you will find something more nuanced and less inevitable. Not everyone is having smaller families; some people are having bigger ones. Not everywhere is welcoming immigrants from different cultures; some countries are trying to restrict their arrival, while others never took to the idea in the first place. Some places are finally starting to see an end to the increase, and even see a decline, in life expectancy. And some major cities have seen outflows of people, which may accelerate in the wake of the Covid-19 pandemic.

## *Towards Tomorrow's People*

My purpose in this book is to demonstrate how population events explain the people of today while also shedding light on what life will be like for the people of tomorrow. Ten numbers will each tell a significant story that speaks to a bigger theme concerning a population trend from around the world: falling infant mortality, growing population, urbanization, declining fertility rates, an ageing population, a rise in the number of the elderly, population decline, ethnic change, rising levels of education and the greater availability of food. Far from being isolated phenomena, these trends are linked in a causal chain. Falling infant mortality gives rise to population growth, which spills into urbanization. Urban people adopt lower fertility patterns, which creates an older population and eventually gives rise to population decline, which invites migration and ethnic change. Meanwhile, the whole system is facilitated by expanding educational opportunities and the increased availability of food.

When these changes are seen together, a single key theme emerges above all others: the move from pre-modernity to post-modernity is, in population terms, a move to greater freedom and control over the most important things in our lives: where and how we and our families live and die. Although the stories and data in this book are a signpost that indicate how things are changing, the future will ultimately be shaped by the decisions of billions of individuals about the most important and intimate things in their lives.

In my last book, *The Human Tide*, I suggested that the future would have three colours: more green (the potential for environmental recovery as population growth slows), more grey (populations getting older) and less white (the impact of ethnic change). These shifts are part of the move to a post-modern demography. The shrinking of populations and the possibility of feeding ourselves more efficiently and with fewer resources means 'more green'; the general rise of ages in the population and the likelihood of extreme ageing in much of the world means 'more grey'; and the boom in African populations while those of European origin fall means 'less white'.

## *A Note on Terms and Data*

In reading this book, it is useful to grasp a few basic terms. The **birth rate** is a measure of births per annum, relative to the population. If the total population were 10 million and 200,000 babies were born in a given year, the birth rate for that year would be 2 per cent or twenty per thousand. The **fertility rate** (sometimes called the total fertility rate, or the TFR) looks at the number of children born per fertile woman in a particular period and calculates how many children the average woman would have if that level of childbearing were representative of her lifetime. For example, if a million women aged between fifteen and forty have 100,000 babies in a given year, they are having, on average, one tenth of a baby each per annum; over a twenty-five-year fertile period, the average woman would have 2.5 children.

Crucially, when demographers refer to fertility, they mean how many children people *actually* have rather than how many children they *could* have. A woman may not have children either because she or her partner has problems relating to their medical fertility or for all sorts of other reasons. If she has no children, demographers would say a woman has a fertility rate of zero, even if she were capable of bearing a large number of offspring.

The **death rate**, or **mortality rate**, is a measure of deaths relative to the population; if the total population was 10 million and 100,000 people died in a single year, the death rate would be 1 per cent or ten per thousand. **Life expectancy**, meanwhile, is a measure of how much longer a person can expect to live at a certain point in their life, based on how many people of different ages are dying in that country at the time. It can be calculated for people of any age, but where age is not specified it refers to life expectancy *at birth*.[16] It is often calculated separately for men and women.

**Median age** is a measure of the age of the people in a society at a point in time. If you lined up the whole population by age, with the youngest at one end and the oldest at the other, the median age of the population would be the age of the person in the middle, located exactly halfway between the oldest and the youngest person in the line.

Data is scattered throughout the book, for it tells a story that cannot be grasped without an understanding of the numbers. But a word of warning: demographic data requires wide-scale collection and verification, and its quality depends on censuses, records and other official data-gathering activities. The data in some places, for some times and for some issues is more reliable than others. In modern societies we take it for granted that a state institution will gather, publish and analyse information on births, deaths and the movement of population across borders, but this is a relatively recent phenomenon. For example, when demographers look at birth rates in eighteenth-century Japan that appear fairly low, they are often unsure whether the numbers are the result of sexual abstinence, abortion or infanticide.[17] A general rule of thumb is that the more recent the data and the more developed the country where it has been gathered, the more reliable it is. So deaths in Finland in 2020 are more certain than migration into Botswana in 1950. Where possible I have used the comprehensive data from the United Nations Population Division.[18] In all other cases, the source is given in an endnote.

Just as the data of the past and present can be unclear, the data that relates to the future is also uncertain. However, while there are no crystal balls, there are issues on which demographers can forecast with confidence. Short of there being some vast calamity, we know how many thirty-year-old Italians there will be in 2050, and we can state with reasonable certainty that South Africans will not be living shorter lives in three decades' time than they are now. But one of the major themes of this book is that nothing is inevitable – how things work out will increasingly depend on the choices people make. In the past we could predict a great deal from the material conditions in which people lived, while economists could forecast how those conditions would develop. With cultural and personal preferences rather than economic factors increasingly shaping demography, it will become harder to forecast.

When you set off on a journey to a distant destination, it could be that before you get there a new road has been built, or perhaps an earthquake will have removed an existing one. However, even if parts of the map that show your journey turn out to be unclear, incomplete or just plain wrong, it makes sense to ensure that when you set out, you have the best map you can get your hands on. By explaining how population is affecting our present, while also setting out the major trends in demography and their implications for politics, economics and society, this book provides a map of the future.

# 1

# Infant Mortality

## *10: The Infant Mortality Rate per Thousand in Peru*[1]

The dusty district of Carabayllo lies between Lima, the capital of Peru, and the Andean sierra. Part of the Incan Empire until its destruction by the Spanish conquistadors in the 1530s, the area was later settled by Spaniards who forced the locals to work on their estates. Despite its location on the outskirts of the capital, it is far from the sort of Third World shanty town so often imagined by Westerners. Its adobe brick old town has some folkloric charm and there are parts of the district where residents have moved closer to the Peruvian middle class. But for the most part it is a scruffy mix of slapdash dwellings and open fields that straddles the town and the countryside, neither fully rural nor fully urban. Makeshift homes lean against the hillside, but then the eye is caught by a modern office block, modest in proportion but suggesting the presence of white-collar jobs.

Carabayllo is not dissimilar from many residential districts in the developing world. The residents of these communities are often progressing from rural destitution towards a standard of living that we in the West have long taken for granted. And an important transition on that journey is the shift from high to low infant mortality, as the death of a baby goes from being a common occurrence to a rarity.

In 1996, a clinic opened in Carabayllo to train community health workers in how to educate local pregnant women in matters like nutrition and hygiene. Mothers who participated in workshops informed those who had not attended the clinic, which helped to spread the reach of this life-saving knowledge. The health advisors came from the communities they served, with understanding of local customs and practices that made them more effective than outsiders. This kind of small-scale initiative made a real difference in the decline of infant mortality.[2]

Sometimes, behaviour can be influenced by simple financial incentives. Early in the twenty-first century, Peru offered prenatal education to pregnant women while giving them cash to attend these classes and to have their children vaccinated. In many cases, it also provided health workers who could speak the local language. Along with on-the-ground initiatives like the clinic at Carabayllo, this has transformed the lives of Peruvian families by reducing infant mortality rates, while also raising the health and life expectancy of Peruvians of all ages and in many regions.

For every thousand babies born in Peru, the World Bank data tells us that ten still do not live to see their first birthday.[3] While other sources vary slightly, one thing is certain: that number is falling fast. In the early 1970s, the infant mortality rate was over one hundred per thousand in Peru, around ten times its current level. Bringing down infant mortality to such a great extent in just a couple of generations is an extraordinary triumph, but the fact that Peru is not particularly renowned for this achievement illustrates how typical it is of many other countries at Peru's stage of development. The decline in infant mortality in Peru in the last half-century has only been a little faster than has been achieved in South America as a whole, while plenty of Asian countries, including China, have a similar track record.

The education of women in Peru has also made a decisive contribution to the huge decline in infant mortality, and not just because it leads to a better understanding of pregnancy, childbirth and childcare. The ability to read and write means that mothers are more able to take control of their families' welfare. As recently as 1970, less than a third of Peruvians enrolled in secondary school; today, almost all girls and boys do.[4] A society in which everyone has access to a basic education is very different from one in which schooling is for a privileged few, not least in regard to child welfare. Women are much more likely to seek and follow medical advice during and after pregnancy and they are also more likely to be able to provide for their children.

We might wonder whether education is the cause of a decline in infant mortality or if the relationship is correlative rather than causative; an improvement in material conditions might explain both fewer deaths and an increase in education. However, statistical analysis proves without doubt that education has a direct impact on infant mortality.[5] Knowledge is the key to extending life, whether that means understanding how mosquito nets can prevent malaria to how using saline and sugar solutions can aid recovery from diarrhoea. Education is also associated with better health and lower mortality in developed countries.[6]

Falling infant mortality is also invariably accompanied by improvements in maternal health. From 2003 to 2013 alone, the rate of women dying in childbirth in Peru fell by more than half.[7]

So the death of infants in Peru has become exceptional, thanks to better and cheaper medicines, higher food standards and cleaner water, in addition to increased access to education. Due to the sheer speed of these improvements, infant mortality has fallen there twice as quickly as it did in European countries. In Britain, it took the first three quarters of the twentieth century to cover the decline Peru has managed in barely twenty-five years. Although Peru is still poor, its rate of infant mortality is equivalent to that in the UK in the late 1970s or Russia at the start of the twenty-first century. To put it another way, a Peruvian newborn baby has half the chance of dying before the age of one as I had when I was born in London in the 1960s.

Just as Carabayllo is typical of much of the developing world, Peru's achievements over the past few decades are typical of many other countries. Global infant mortality rates halved between the early 1950s and the early 1980s, and have halved again since then. More progress has been made in this area in the past few decades than in the whole of previous human history.

### *Infant Mortality: What It Is and What Drives It?*

The fact that we now measure infant mortality per thousand is significant. It was once measured per hundred, because in most pre-modern societies, around one in three children would die before they were a year old. The great change in lifestyles and living standards of the last two centuries in some parts of the world, and of the last few decades in others, have dramatically improved the life chances of the young; these days, the highest-achieving countries, like Japan, manage around *two per thousand.*

These days, we associate death with the elderly, but in pre-modern societies the very young were at greatest risk. For Christians who believed you needed to be baptized to get to heaven, making sure your children underwent this rite as early as possible was a priority. In the Italian city of Padua in 1816, when almost 15 per cent of babies died in their first six days of life, three quarters of children were baptized before they were two days old. By 1870, the chance of such an early death had halved, and so had such early baptisms.[8] One likely explanation is that parents knew they had time to ensure that their offspring would not be cast into limbo and could be a little more relaxed.

Where mothers are illiterate and live on poor diets in unhygienic conditions, babies struggle to make it through the first year. In the past, this was true for just about everyone. King Henry VIII's six wives only produced three surviving infants between them, despite their access to every sixteenth-century comfort, luxury and attention that money could buy.[9] Those three children between them produced no heirs. It did not help that two of Henry's wives were executed, one of whom was aged just nineteen, with many potential years of childbearing ahead of her, but the other four died of natural causes. In the UK we all remember that Henry had six wives, but the fact that he had no grandchildren – or at least no legitimate ones – is far less recalled. Six wives and not a single grandchild gives us a sense of how often bloodlines were closed off, even among those at the very top of society.

I sometimes ask people when they think an eldest son last inherited the English throne from his father and passed it on in turn to his son. The answer is more than six hundred years ago, and the king in question was Henry V. Sometimes a smooth succession did not happen because of political events and usurpers, but it often had to do with the early death of the heir. Henry VIII's own older brother, Arthur, died before their father, Henry VII. George II's son Frederick died before his father, with the throne passing to his own son, who became George III. And George V only inherited the throne because his brother Albert Victor died in his late twenties, before their father had even ascended the throne.

The prolific loss of children explains why the population grew so slowly in the past, and why there were barely a billion people on the planet a couple of centuries ago, compared to seven billion now. Things did not improve much between the reign of Henry VIII and that of Queen Anne a century and a half later; she had seventeen pregnancies and no surviving offspring. The fate of her dynasty was sealed when her son William, Duke of Gloucester died after eleven sickly years of life. Like her sister Queen Mary II, her bloodline closed off. With the demise of the Tudors and Stuarts, at least of Protestant ones, a second cousin from Hanover was brought in to fill the gap in 1714. Royal families are useful as examples because we know so much about them, but what is true of them is true also of the less privileged in society. The family trees of the past would be full of dead branches like those of Henry VIII and his wives, people who left no living descendants. One thing that we all share, whether we're royal or commoner, is that every one of our ancestors successfully reproduced.

The probability of death in any year of life tends to decrease over the period to adolescence and then to rise as we approach old age. There was a time when, in order to compensate for the death of those who died before having children themselves, women needed to have an average of six births or more to keep the population stable. That number is now generally considered to be 2.1.

## *The Shadow of Death*

The inescapability of death has haunted humans ever since we first became conscious. Images of gloomy subterranean rivers and fiery furnaces terrified our ancestors and continue to terrify many of us. It has shaped our religions, our mythologies and our art. Vast resources are ploughed into medicine and healthcare in order to delay it. What else is Britain's National Health Service and the American healthcare industry but a huge collective effort to put off dying? Humans have always lived with the strangeness of knowing that our family and friends will one day disappear forever, and that the same fate will be ours too. For demographers, death, or 'mortality', the statistical tendency to die, is one of the basic givens of the discipline.[10]

You might think of survival as a race in which a hurdle has to be jumped every year.[11] The early hurdles were once very high, and many people fell before clearing the first few. Demographically, humankind was living close to a state of nature, and nature is prodigiously wasteful of life. From only a very small number of figs will even one of many seeds become a fig tree and produce its own figs. Almost a quarter of baboons born in the wild die before they are a year old, which is fairly typical for primates.[12] Infant mortality rates for humans in pre-modern agricultural societies were similar to, and sometimes higher, than those of apes. Things were little better for our hunter-gatherer ancestors, a quarter of whom failed to reach their first birthday, while another quarter didn't live long enough to reproduce just about everywhere until the nineteenth century. Had it been otherwise, the huge recent growth in human population would have occurred much earlier.

Today our hurdles, and particularly the early ones, are much lower, and so only a very small proportion of those born in the developed world fail to clear them. The vast majority clear enough hurdles to be able to procreate themselves. And while every early death is a tragedy for that young person and their family, the fact that these events are such a rarity in so much of the world today is a cause for celebration.

Even today, many births in sub-Saharan Africa take place far away from towns or bureaucratic activity, so go undocumented, and the same is true of the more remote parts of Asia. Joe, from Chiang Rai in northern Thailand, did not receive his citizenship and ID card until his late teens, since his parents neither understood how nor were prepared to register his birth. This causes headaches for demographers, who rely on accurate statistics, and more importantly can lead to inconvenience for someone from a remote background who wants to integrate into modern life. 'I couldn't go to places I wanted to go,' Joe told a British journalist. 'Police always stopped me from leaving my village, as I didn't have anything to prove I was Thai.'[13] If births – and deaths – go unregistered, infant mortality rates are often based on estimates rather than hard facts.

The data can also be distorted by the fact that the infant mortality rate is likely to be lowest in the places where it can best be sampled – these are the same places where facilities are available to prevent infant death and where people are likely to have the highest level of education. While adjustments can take into account the fact that urban settings provide the best healthcare and the best places to count, they can only be approximations in countries where most people live beyond the reach of doctors, let alone demographers.

Another limit on the data is the fact that some social change happens so quickly that by the time the data has been analysed the reality has changed, and few population measures can change as quickly as infant mortality. In bad times, a plague or famine can sweep away a whole generation. The very young, as well as the very old, are generally more vulnerable to waves of disease; in recent times, the arrival of a modern innovation, like a vaccination programme or basic maternal education, can cause death rates to tumble for the young in particular but also for the population as a whole. The chlorination of the water supply of the US, for example, more than halved deaths from typhoid in just a dozen years, before eventually more or less eliminating it completely.[14]

Although there may be glitches in the data and some uncertainties, falling infant mortality is one of the most positive – and powerful – demographic trends of the contemporary world. However, despite a picture that is either good or rapidly improving almost everywhere, it is far from even.

### *Leaders and Laggards*

The fastest progress in bringing down infant mortality rates in the last fifty years has come from the developing world. On the whole, infant mortality declines with a country's wealth and level of education, but child health is being prioritized by the governments of poorer countries and international aid agencies, and the welfare of their children is the highest priority of most parents. As a result, relatively poor countries are narrowing the gap more quickly than might have been expected, as we have already seen in Peru.

The United States is among the rich, developed countries with a relatively low – and falling – infant mortality rate of a little above 5.5 per thousand.[15] While this is much lower than developing countries like Peru, it remains disappointingly high for a country that is not only extremely wealthy but is also at the centre of advances in global medical research. The relatively poor performance of the world's richest country on this measure reflects the limited healthcare that is available to poor people. But even in the US, infant mortality has halved since the early 1980s. There is no doubt that it is harder to make progress when the level is already low, but other countries have achieved it. Forty years ago, levels of infant mortality in Western Europe and the US were about the same; today, the rate in France, Germany and their neighbours is half that of the US. Eastern Europe, where the infant mortality rate was more than double that of the US in the early 1980s, today has a similar one.

Delving into population data gives us useful insights, so it's worth looking a little more deeply at why the United States has in recent years become a relative laggard among leading countries.

As in so many other respects in the US, race is a major contributing factor to infant mortality. The rate suffered by African Americans is more than twice the level of infant death experienced by society as a whole, but Hispanics, who are still relatively poor, have a slightly lower rate than the average. The best rate is achieved by Asian Americans, usually recent immigrants or their children, for whom an infant mortality rate of between three and four per thousand is not far from the world's top achievers.[16] White Americans do slightly better than Hispanics as a whole, but significantly worse than those from Cuba and Central and South America, and much worse than Asians.[17]

Nevertheless, the story of African American infant mortality brings good news. A 24 per cent decrease in the number of deaths of black infants was achieved in a single year in Cincinnati through the empowerment and education of local women. This included an increase in the number of community health workers and prenatal groups for pregnant women.[18]

Infant deaths are most likely to occur among the youngest and the oldest mothers (under twenty and over forty). In the case of the former this tends to be about these mothers living in relative deprivation, often in difficult circumstances and with little education. In the case of the latter it is mainly about biology: the further into middle age women have children, the more difficult pregnancy and childbirth are likely to be. This gives us some insight into where the worst problems lie, and where any US administration should focus its efforts, but there is no getting away from the fact that the overall poor performance of the US has to do with a lack of affordable healthcare provision.

By geography, it is the poor states in the American south-east that fare the worst, and these are the states with the highest proportion of African Americans. But materially deprived and overwhelmingly white West Virginia, famously the home of hillbilly culture and a region wracked by post-industrial blight, has a similar rate of infant mortality to Alabama or Georgia.[19]

In Japan, South Korea, Singapore and Norway, the infant mortality rate is just two per thousand. This has also been achieved in Estonia, one of the most advanced parts of the former Soviet Union and a country that only relatively recently emerged from Communism. Swathes of the developed world have an infant mortality rate of three per thousand; there is no reason why the highest achievers will not continue along this path – before too long, a day may come when infant mortality is measured, like maternal death in childbirth, not per thousand but per ten thousand or even per hundred thousand.

When it comes to *reductions* in infant mortality, the star performers are even more impressive than Peru. In the Maldives, thanks in part to rapid economic progress spurred by a booming tourist sector, the infant death rate has fallen by a staggering 85 per cent since the early 1990s. As recently as the 1960s almost a quarter of children in the Maldives died before the age of one. Since then it has fallen to seven per thousand, less than 1 per cent. The extent to which infant life is now valued there is illustrated by the fact that in early 2019, a critically ill child was winched into a helicopter by the Indian Coast Guard and rushed to hospital in Male, the capital.[20] This gargantuan effort to save a single life would have been unimaginable in previous generations, when the resources were not available to hold life dear. The Maldives, a still-developing country, has achieved survival rates that are only slightly behind the laggards of the developed world such as the United States, and equal to the poorer countries of the European Union such as Romania.

## *No Place for Complacency*

While it is true that in absolute terms, the infant mortality rate in the United Kingdom is low, at less than one seventh of the global average, there are worrying signs. There is no doubt that it has improved over the long term – it is less than half what it was in the late 1980s, and a fifth of what it was in the mid-1960s – but the improvement seems to be coming to an end; there has, in fact, been a modest and short-lived reversal. A rise from 3.6 per thousand in 2014 to 3.9 in 2017 may not seem more than a statistical blip, but any rise at all defies what we have come to think of as inexorable, one-way process.[21] As we would expect, the figures were worst for the most deprived areas.[22] UK infant mortality is still only around two-thirds of that experienced in the US, but it is almost double the top performers, such as Finland. More recent data – close to 3.5 in 2020 – shows a promising reversion.[23]

The causes of this regression – or at least standstill – in the UK are not entirely clear. Opposition politicians will blame government-sanctioned 'austerity' over the past decade, but more money is being spent on healthcare than ever before. The factors we associate with low infant mortality, such as female education, have continued to improve – each cohort of mothers has more schooling and university education than the last. On the other hand, the cohort of childbearing women in the UK looks increasingly different from that of a decade or two ago. Nearly 30 per cent of UK babies are born to foreign-born mothers, well over twice the number in the early 1990s. It is true that many of these mothers come from countries where infant mortality rates are lower than in the UK – Poland, for example, which is the most common origin of foreign-born mothers – but almost as many come from Pakistan, where infant mortality is much higher. Despite the best efforts of the National Health Service, it is perhaps not surprising that the entire gap between Pakistani and UK levels of infant mortality has not been spanned in a single generation. Rising levels of obesity and diabetes among the general population and their impact on maternal health during pregnancy do not help, either. Another possible factor is the rising number of older mothers for whom childbirth is more biologically challenging. In the decade to 2018, the number of children born to women over forty-five in England rose by 46 per cent.[24]

There is one possible reason why a rise in foreign-born mothers is linked to flattening or even rising infant mortality. Mothers born in countries like Pakistan may struggle to navigate their way around the NHS and to access other social services, given possible cultural and linguistic barriers. A child's health is to an extent determined by its mother's education, health and nutritional history. Women who have come to the UK from poor countries, besides being less educated, are likely to have experienced worse conditions in their childhood and adolescence, and there will be a knock-on effect when they reach motherhood. Regardless of how much effort is taken to counteract such effects, even a small residue will shift the dial on infant mortality slightly upwards.

Despite the discouraging data from the UK, we have seen significant falls in infant mortality rates even in countries where it was already low; globally, it is between a fifth and a quarter of where it was in the middle of the twentieth century. Despite this improvement, we should not rest on our laurels; a child dies every five seconds somewhere in the world, and the worst-performing countries have a level of infant death that is forty times worse than the best. The global annual number of deaths of children under fifteen was more than six million as recently as 2018, with more than five million of those aged under five.[25] We can expect this to decrease as improvements in education, living conditions and medical technology continue to spread. As one American mother told how one of her own children had narrowly avoided death: 'Human progress saved my baby, and will save many more.'[26]

Over half the world's infant deaths occur in Africa.[27] In Sierra Leone the infant mortality rate is still around eighty deaths per thousand. Malaria, pneumonia and diarrhoea are widespread and the country also has the world's highest prevalence of maternal death in childbirth. Yet even there, great strides are being made and the rate of infant mortality is half what it was in the mid-1990s.

While the situation is worst in sub-Saharan Africa, there are poorly performing countries all over the world. There is no inherent reason why infant mortality should be twice as high in Pakistan as in India; in the 1970s, these two South Asian rivals were at almost the same point. But while the Indian rate has fallen by three quarters since then, the Pakistani rate has only halved. Pakistan has not been helped by factors such as the prevalence of conspiracy theories that reduce vaccination uptake by children. 'The Hindus are lacing the vaccines with pigs' blood to send us to hell,' a reluctant parent explained to one health worker.[28] In 2019, a health worker and two police guards were shot dead following a social media health scare, and polio vaccination had to be suspended.[29] Ignorance continues to cost lives.

### *Inequality, Virtuous Circles and the End of Low-Hanging Fruit*

Recent decades have seen rapid economic growth in the developing world while median wages in the developed world have stagnated. This has led to economic convergence between nations, with the poorest nations closing the gap, while the economic inequality *within* nations has increased. For example, in poorer countries, a growing middle class has diverged from the local poor, while wealthier European, North American and Asian economies have seen the rise of a super-rich elite. The same patterns have been observed in infant mortality. The worst countries have made the fastest progress, so the gap is closing at the international level. For example, the infant mortality rate in Malawi in 1950 was almost 150 per thousand greater than that in the US; today it is just thirty-five per thousand greater. On the other hand, there are growing disparities *within* Malawi, where progress has been uneven. The best results are achieved in urban areas, where facilities and education are easiest to supply, while rural areas are relatively neglected. Corruption, shown in the preferential treatment of certain regions and in the creation of a middle class, also exacerbates divergence within borders. In Malawi, infant mortality in wealthy urban areas was, according to a 2014 survey, less than half the level in poor rural regions.[30]

As risks fall, so people focus on preventing them – and this is not only true in demography. For example, since the early 1980s, deaths in fires in England have more or less halved. This is thanks to the reduction in domestic open fires, an increase in fire-resistant furniture and other measures. However, when they do happen, they are more shocking; the Grenfell Tower fire in June 2017, from which seventy-two people died, rocked the nation. When a disaster strikes, the more unusual it is, the more it impinges on the national consciousness. But the more impact it has, the more countermeasures are taken, so the level of deaths continues to fall as a result.[31]

The analogy with infant mortality is clear. An aid worker who once lived in one of Africa's poorer countries told me that infant death was so common fifteen or so years ago that an employee of his might not take a day's leave if it happened to them – it was accepted as part of life, and so less was done to resist it.[32]

As infant mortality rates fall, each occurrence has greater shock value and society determines to do more to combat it, which in turn further reduces mortality rates. The recent modest rise in the UK is leading to investigations that will likely influence policy and practice, which in time will reverse the negative trend.

On the other hand, there is a concern that the easiest gains in infant mortality may already have been garnered; having plucked the low-hanging fruit, future gains will be harder to achieve. The 'easiest wins' are in the latter eleven months of the first year of a child's life, when such initiatives as vaccination programmes can be introduced. Success in the vulnerable first month is much harder to gain, as it depends on intervention prior to and during childbirth, in addition to more complex issues such as maternal nutrition and health. In some cases, deaths in the first month, when the child is most vulnerable, have resulted in such high infant mortality that rituals and rites have been developed to cope with it. 'When I was working in Uganda, I came across a ceremony to mark a child reaching its first month,' one aid worker told me. 'It was as if the child wasn't fully considered human until that point, and its death was more like a miscarriage or stillbirth.'[33]

*Mothers Matter*

There was a time when childbirth was a great threat to a woman's life. Along with infant mortality, maternal mortality was the main pre-modern constraint on population growth. If a girl survived her first year and then made it to childbearing age, she may well have died during her first or second pregnancy, or while giving birth. The stepmother was once a far more significant feature in everyday life than is the case today. Many men found themselves left with children to raise and no wife, and so needed to find a replacement.

Pregnancy and childbearing can be hazardous; high blood pressure is common in the former and bleeding and infections often occur during the latter. Furthermore, deaths often occur during abortions, especially when these are unsupervised and take place illegally. All these situations are more likely to prevail when a mother is giving birth in poverty, when she has a low level of education and where basic facilities are either distant or unavailable.

Like infant mortality, maternal death in childbirth is still common in parts of the world, but these areas are shrinking and it fell by more than a third in the years from 2000 to 2017 alone.[34] The retreat of this particular form of human misery has much in common with the reduction in infant mortality, and as with infant mortality, the provision of professional care can make a big difference.

The story of a midwife in a working-class district of Colombo, the Sri Lankan capital, is illustrative. Ariyaseeli Gunaweera shows a new father how to create and manipulate a simple DIY breast pump, which helps to feed his three-day-old baby and ease the discomfort of its young mother's breasts. More than 90 per cent of Sri Lankan mothers receive such visits in the days following childbirth, and they are often a life-saver for both mother and baby. 'No one else will come to help us,' the grateful grandmother tells a reporter. 'Only the midwives.'[35] A breast pump costs a tiny amount to produce, but along with instruction on its safe, hygienic use, it can save lives every day.

The success of the Sri Lankan system is a result of carefully kept registers that track the progress of pregnant women and highlight those who are judged to be at risk. Sri Lankan midwives form close relationships with mothers-to-be, which allows them to have frank conversations about issues like sex and domestic violence. 'I'm part of the family now,' says Ariyaseeli. Sri Lanka has become a model for mother and baby care, and delegations from across South Asia and beyond travel there to study its midwifery services.

With its bloody civil war still within recent memory, it is often forgotten that Sri Lanka was something of a model colony within the British Empire until it gained independence in 1948. With a representative council since the late nineteenth century and universal suffrage to a legislative assembly since the early 1930s, it was relatively advanced constitutionally and also benefited from a tea-exporting industry that was integrated into the global economy. Training for midwives had begun way back in the 1880s. In the year 2000, Sri Lanka's maternal death rate was fifty-six per hundred thousand live births, a remarkable exception among developing countries but much higher than in the developed world. However, following the development of new practices in postnatal care, Sri Lanka's maternal mortality rate had fallen to thirty-six per hundred thousand by 2017.[36] There is still a considerable gap with the developed world – avoidable deaths and abandoned orphans remain a distressing reality – but in many parts of the world such cases more than halved in a generation.[37] Rates of maternal death are higher in rural areas, where mothers are less accessible to midwives. When access to services and facilities makes a difference, it is usually better to be in a town or city, where these can more easily be accessed.

As with infant mortality, simple interventions can make a significant difference. Studies show that organizing groups of women to discuss pregnancy and childbirth, and to share basic healthcare lessons, can be an effective way of improving their health and reducing their mortality rates.[38]

Elsewhere in South Asia, Afghanistan was, at least until recently, also making progress in this area – since 1990, its maternal death rate has more than halved, but there is a long way to go before it reaches the standards of countries with long traditions of excellent mother and baby care. And, as ever in a country like Afghanistan, the data is uncertain. Recent findings suggest that the improvement has been smaller than previously thought, and that although the provision of midwives has improved, they reach fewer mothers in rural areas. In fact, Afghanistan appears to be the only country outside sub-Saharan Africa where more than one childbearing mother out of a hundred dies.[39]

Maternal mortality rates matter both in themselves and for what they represent. A high maternal mortality rate is usually associated with a high infant mortality rate; both are rooted in a lack of education, hygiene, medical facilities and public health. Falling maternal mortality rates illustrate the importance of female education and psychological support for new mothers, just as falling infant mortality rates do. In traditional rural societies, childbirth is especially risky for the mother. Deaths of mothers in childbirth in England and Wales fell from between forty and sixty per thousand births in the nineteenth century to around forty-two per thousand by 1930. Today, the figure is around seven per *hundred* thousand.[40]

While the global decline in maternal death has been an

outstanding achievement of the post-war world, there are some unfortunate exceptions. Here too, the United States lags behind, again owing to the relative inequality of care and the patchy provision of health services to the poorest in society. Deaths of American women in or shortly after childbirth are at around nineteen per hundred thousand births, which is not only high compared to other developed countries but worse than a generation ago, and at least double what it was in the late 1980s.[41] And it is not just those in poverty who are suffering. Celebrities such as Serena Williams and Beyoncé have spoken of how they came close to death while giving birth. 'I almost died after giving birth to my daughter, Olympia,' wrote Williams. 'First my C-section wound popped open due to the intense coughing I endured as a result of [my] embolism. I returned to surgery, where the doctors found a large hematoma, a swelling of clotted blood, in my abdomen. And then I returned to the operating room for a procedure that prevents clots from traveling to my lungs.' Williams is grateful for the medical care she received, but not all women are so fortunate.[42]

The United States has been remarkably slow in developing the postnatal supervision that has made all the difference in countries like Sri Lanka, and the issue of race once again raises its head. Black women are more than three times as likely to die of complications in childbirth as white women, and almost four times as likely as Hispanic women. The issue was introduced into the political arena by erstwhile Democrat presidential candidate, now vice president, Kamala Harris, who insisted it was the result of racism in the healthcare system. But if that is the most pertinent explanation, it is hard to understand why Hispanics outperform whites on this score.[43] As with infant mortality, these matters of life and death require careful examination and analysis, followed by carefully planned and assiduously executed policies.

As with infant mortality, the most critical time for maternal

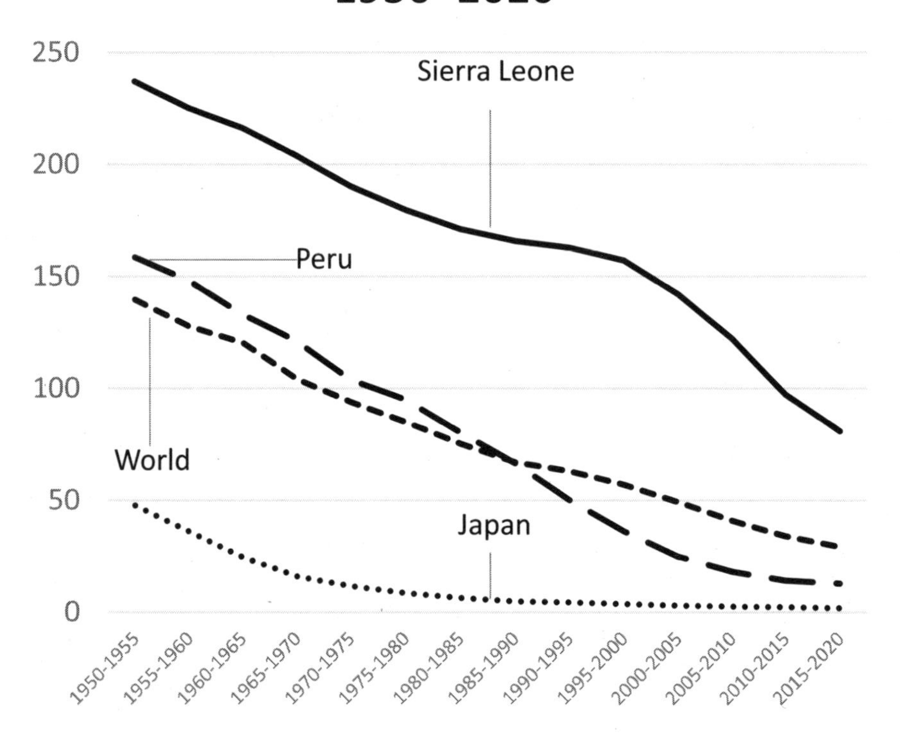


Source: United Nations Population Division

Infant mortality has been falling fast, and it has declined quickest in countries where it is highest; in Sierra Leone, the number of children who have died in their first year of life has fallen from around a quarter in the early 1950s to much less than a tenth today. In Japan, progress is harder to see because infant mortality has been so low for so long, but at barely two deaths per thousand children, the Japanese rate is among the best in the world. The global picture is improving, and educated parents – especially mothers – as well as rising incomes and increased access to healthcare and maternity services have all contributed to this success story.

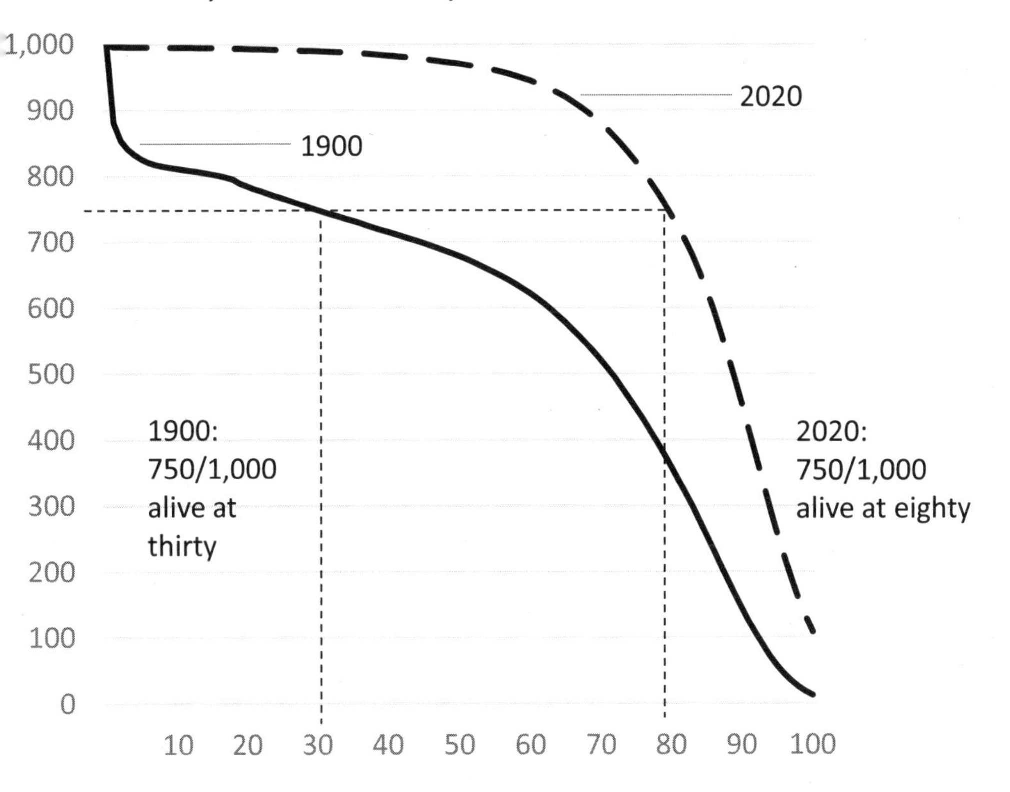


Source: US Social Security

If we take women's survival rates by age in the US in 1900 and assume them to apply to today's newly born cohort as they progress through life, we could expect only 800 out of 1,000 still to be alive at the age of ten, and around two-thirds to be alive at the age of fifty. By contrast, using the mortality rates that prevailed in 2020, more than 99 per cent would be alive at ten years old, and 97 per cent would still be alive at the age of fifty.

Using the same data, a quarter of the 1900 cohort would have died by the time they reached the age of thirty; for the 2020 cohort, it would take until the age of eighty for only three-quarters to remain alive. Undertaking the same exercise for the cohorts born between these two years would produce a series of gradually shifting curves, with a higher share of the cohort surviving at any given age.

health is immediately following childbirth. In the US, which is in this respect not atypical, slightly more women die while giving birth or in the week thereafter than in the subsequent fifty-one weeks.[44]

### *The Difference that Infant and Maternal Mortality Make*

A significant decrease in infant mortality profoundly changes a society. In a developing country, the population gets younger – the median age in the French Indian Ocean territory of Mayotte, for example, fell by fifteen years between 1950 and 1985, as infant mortality plummeted by around 90 per cent. As a result, more schools needed to be built, more people were required in child-oriented professions such as teaching and the economy had to absorb a burgeoning young workforce. Like the challenges involved in caring for an ever-increasing number of elderly people, these demands result from positive demographic news – a push-back against death. It may be a worthy objective for a country like Japan to bring down its infant mortality rate from two or three per thousand to one or less, but infant mortality has long been at such a low level there that further falls will make no material demographic difference.

Even in a country like Peru, eliminating infant mortality altogether would not significantly impact population size, because only around 1 per cent of babies die. *Falling* infant mortality rates were once the great engine of global population growth. Now it is *low* mortality, where it is coupled with persistently high fertility, that fuels population explosions.

In time, lower infant mortality leads to lower fertility rates, partly because parents who expect to lose fewer children have fewer children, and partly because of the association between smaller families and prosperous lifestyles. Initially the increase in survival rate brings down the median age in society, but once family size declines, the large cohort who survive are succeeded by smaller ones, as happened to the baby boomers in the West after the Second World War.

Cynics might suggest that those people who are worried about population growth should avoid assisting less developed countries to reduce their infant mortality rates, but there is a long and ugly background to such ideas. Not only is such an approach lacking in compassion but it is impractical to imagine that we could hold millions of people for ever in some pre-modern state of misery, even if we were callous enough to want to do so. Instead, for those wishing to lower the rate of global population growth, the most humane and practical way of doing so would be to hurry those places in the early stages of the demographic transition through its subsequent stages as rapidly as possible, to a point of prosperity, female emancipation and choice. For this to happen, societies that are the most materially and educationally limited will have to pass through the demographic phase when mortality plummets but fertility remains high before they can get to the subsequent phase when birth rates fall.

Despite the relative shortcomings of America and Pakistan, we should not lose sight of how far we have come in reducing infant and maternal mortality. There is every reason to believe that in the years to come, increasing postnatal support in developing countries will make a baby losing its mother a rare tragedy. Short of there being a global calamity, reverses such as those in the UK are unlikely to become statistically significant. It may be that the UK experience simply represents the difficulty for very good performers to reach the ranks of the very best.

We know what the future looks like with regard to low infant and maternal mortality, because that is already the situation in much of the developed world. Some people think death might be banished entirely, but for now this idea remains on the wilder shores of science fiction. We will examine such seemingly outlandish theories later, but in the sober realm of the realistic, while infant mortality and death in childbirth will not be abolished altogether over the next few decades, very low levels of both will continue to spread around the world. And although a reduction from two per thousand to two per ten thousand or even two per hundred thousand is a worthy aspiration, rates of infant mortality and death in childbirth will be so low to start with that further improvement will barely make a difference to the numbers.

Much of the world is still moving towards that condition, and when the deaths of infants and mothers plunge, the population initially grows at breakneck speed. This is the expansionary stage of the demographic transition that Europe started to leave behind more than a century ago and which much of the world is currently passing through. But in Africa it's still got decades to run – and it will change *everything*.

# 2

# Population Growth

## *4 Billion: The Population of Africa by 2100*[1]

Malngaye Adam has ten children with his wife Kattouma, whom he married when she was fifteen. Interviewed by a *Financial Times* journalist, he admitted that had she not been sufficiently fecund, he would have found a second spouse. They live in a rural backwater in southern Chad, at constant risk of drought and raids from Islamist terrorists, and this patriarch has no regrets about his many offspring: 'It's a matter of pride to have a big family. Lots of children help you. It was not my choice. God gave them to me.' In countries like Chad, people's ability to shape their own lives is limited by traditional family structures and attitudes – many children tend to arrive, whether or not they are wanted.[2]

The stories of people like Malngaye and his family illustrate why the population of sub-Saharan Africa is increasing so quickly. According to the best estimates of the United Nations, the continent will be home to around four billion people by the end of the twenty-first century. Africa's great population expansion is one of the most important trends on the planet today; it also has the potential to transform our global politics, international relations, economics, culture and ecology. It's being driven, like the other great population expansions of the last two centuries, by two factors: a combination of ongoing high fertility and falling mortality, particularly among the young, that we observed in the last chapter.

In the past, an endless flow of new people turned up at the railway station of life, but they were spirited off by the train of death almost as quickly as they arrived. Now, the people continue to arrive in huge numbers but departures are much reduced, so the platforms are thronged. This trend began in nineteenth-century Britain. It went global in the twentieth century and is now happening in Africa, the final frontier of the demographic transition.

Contrast Malngaye Adam with Henry VIII, whose lack of reproductive success we discussed in the last chapter. It is surprising that the former, living at the bottom of the global social pyramid, should be so successful in rearing a family while the latter, very definitely at the top, failed to produce the heir he so desired. Of course, there have always been those among the poor who have had procreative good fortune and those among society's most privileged who have had procreative bad luck, but the ages in which and Malngaye Adam and Henry VIII lived most powerfully explains the difference in their reproductive fortunes. In a pre-modern world, Malngaye Adam would not have expected all ten of his offspring to survive to maturity – such an outcome would have been a happy fluke. Today, on the other hand, even in a country as poor as Chad, the survival of all his children beyond infancy is fortunate, but not exceptional. As recently as 1950, almost one in five babies in Chad died before the age of one. Today it is less than one in ten, fifty times worse than Japan's infant mortality rate but moving in the right direction.

Chad has all the characteristics that encourage high fertility rates. Less educated women have more children than more educated ones; with only around one in five young Chadian women able to read, and just 12 per cent starting secondary school, it is unsurprising that fertility rates are still high.[3] The country is in the first stirrings of demographic transition; things have improved enough to bring down mortality rates but fertility rates are as yet unchanged. As would be expected, Chad has a very young population; the median age is just sixteen, half a decade younger than it was in the middle of the twentieth century, simply because more of its young people survive.

Life continues to be tough for most people in Chad, and it would be naive to think that nothing could bring down the population size. Many people live precariously, with insufficient nutrition for them to thrive. It has been estimated that up to three million people are facing food insecurity because of the intensified cropping, deforestation and overgrazing in the Lake Chad basin.[4] At one point it was thought that global warming was causing Lake Chad to shrink, though the lake has expanded rapidly in recent decades and climate change is now blamed for the excessive rainfall in that part of Africa.[5] Whether there is too much water or too little, the region's population continues to grow exponentially.

Niger, next door to Chad, is another vast country with a small but rapidly growing population. Relatively empty, it covers an area more than five times as big as the UK, but its population is not much more than a third the size.[6] Niger's National Hospital treats many desperately malnourished children. 'I have six children and two have suffered from acute malnutrition. It's because I don't produce enough milk because I don't have enough to eat,' lamented Amina Chaibou.[7] This young mother's pitiful situation should not blind us to the fact that those people who have been admitted to the hospital are the most fortunate. Many others suffer unseen, far away from the resources of the big city. But neither should it obscure the fact that women like Amina Chaibou are starting to become the exception in Africa rather than the rule. If that was not the case, Niger's population could not have doubled in the first two decades of the current century, and nor could its infant mortality rate have halved. For much of human history, the otherwise unstoppable force of population growth was limited by the resources available. This, in a nutshell, was the argument of Thomas Malthus, the founder of modern demography. And now that the resources available are much less of an immediate problem for many, population growth can go marching on.

It may be that population is about to meet a global disaster and get knocked back, finally proving Malthus right after two hundred years. In the wake of Covid-19 we are certainly more aware of the destructive potential of pandemics. But for the moment, the population of Chad and its neighbours continues to boom. And just as China has been rocking the world's economic foundations in recent decades, Africa is set to transform our global demography. The great economic shift has taken place in the east, and the great demographic shift will occur in the south.

### *Africa: Population Explosion*

Chad is one of Africa's biggest countries in terms of area, being larger than the UK, France and Germany combined, but it is also one of the world's poorest. Its population is doubling every couple of decades; eight million became sixteen million in the first two decades of the twenty-first century, a result of an excess of births over deaths rather than of immigration. Sub-Saharan Africa consists of around fifty countries, enormously varied in their geography, climate, ethnicity, history, religion and resources. Zambia is a world away from South Sudan, as Rwanda is from Guinea. N'djamena, the capital of Chad, is closer to Paris than to Cape Town. The distance from Dakar, the capital of Senegal, to Mogadishu, the capital of Somalia, is more than double the distance from Dakar to Madrid. Nevertheless, some generalizations can be made about African demography, just as they can about Europe or Latin America.

For centuries, Africa was a source of slaves, first for Muslim traffickers and later for the Europeans who exported people to the Americas. Arab traders are believed to have taken as many as fourteen million people and Europeans around twelve million.[8] Centuries of rapacious slave-raiding and colonialism combined with Africa's pre-modern demography meant that the continent was, until recently, hugely underpopulated relative to its size. Africa is several times bigger than Europe, but in 1950 it was home to fewer than two hundred million people, less than half Europe's population at the time. For decades, high fertility rates and high mortality rates cancelled each other out, resulting in a stable population. Now, with survival rates increasing, Africa is breaking out of demographic obscurity.

The degree to which the demographic tables have been turned is illustrated by the memoirs of Captain James Frederic Elton, a Briton who travelled in coastal East Africa in the 1870s. At that time, Britain was using its population surplus to settle its colonial possessions, and Elton was tasked with appraising the region's potential for settlement. In his journal he recorded his hope that despite the apparent scarcity of native people, the indigenous Africans would not disappear entirely 'as the white man advances', an occurrence that had 'sadly marred many pages in the history of our colonisation'. In the nineteenth century, Europeans were thrusting forward everywhere and non-Europeans were in retreat,[9] and Elton was unusual in lamenting the disappearance of non-European races before their all-conquering masters. Although the advance of 'white man' and his displacement of 'the natives' seemed inevitable, some people still found it regrettable.

Such attitudes toward indigenous peoples were hardly unique to Africa. In the nineteenth century an Australian newspaper characterized what was a fairly common attitude when it declared that 'when savages are pitted against civilization, they must go to the wall; it is the fate of their race. Much as we may deplore the necessity for such a state of things, it is absolutely necessary, in order that the onward march of civilization may not be arrested by the antagonism of the aboriginals.'[10] In *The Descent of Man*, first published in 1871, Charles Darwin forecast that 'At some future period . . . the civilized races of man will almost certainly exterminate, and replace, the savage races throughout the world.'[11] In the middle of the nineteenth century, when the United States took over what had been the northern half of Mexico and what is now the American west, one US senator predicted that the Mexicans would 'perish under, if they do not recede before, the influences of civilization . . . They are destined . . . to give way to a stronger race.'[12] Racial attitudes that can now be framed as 'Darwinian' were, in fact, common several decades before Darwin ever put pen to paper. Darwinism may have been misused to provide a pseudo-scientific basis for racism, but it hardly created it.

But what seemed in Darwin's and Elton's day to be a permanent feature of the present and future has turned out to be nothing of the kind. According to recent studies by the UN, Africa will by 2100 be home to twenty times the number of people who lived there in 1950. At four billion, Africans will outnumber Europeans by six to one. In 1950, one person in every fourteen lived in sub-Saharan Africa; by 2100, it will be one in three. At a global level, it will be the fastest shift in the relative population sizes of different ethnic groups since the conquest of the Americas in the sixteenth century led to the collapse in their indigenous population.

The rest of the twenty-first century will see a steady rise in African life expectancy and a corresponding fall in mortality rates, particularly among children and new mothers. We can also expect significant migration, both within Africa and to Europe, although much will depend on the pace of economic and political development of the former and the attitude to immigration of the latter. Inevitably, Africa's progress and its ability to absorb its own population growth will be patchy, but however things turn out, the continent is set to be the home of tomorrow's people. If the 'out of Africa' hypothesis is correct and humankind started in Africa, we are returning to our roots.

### *Out of Africa*

'It is unacceptable that sometimes in certain parts of Milan there is such a presence of non-Italians that instead of thinking you are in an Italian or European city, you think you are in an African city . . . Some people want a multicolored and multiethnic society. We do not share this opinion.' These were the sentiments not of a random passer-by or a far-right sympathizer in northern Italy, but of the Italian prime minister Silvio Berlusconi in 2009.[13]

We will look later at the impact that large-scale immigration is having on the developed world, but let us first consider the population explosion in parts of the developing world that is fuelling it. In Asia and Latin America, population growth is slowing: in most countries the median age is rising and the potential for mass migration is reducing. This is why, in recent years, the net flow of Mexicans into the United States has reversed. However, Africa's great growth spurt lies in the future. Much of the migration will take place *within* Africa, from the countryside to towns and from less successful to more successful countries, but much will also be heading towards Europe, the most accessible part of the developed world. It is also the continent that Africans are most familiar with because of their colonial history and knowledge of languages, especially English and French. A recent graduate in Senegal, frustrated by the lack of jobs at home, is more familiar with Paris than Beijing; their equivalent in Lagos knows more about London than Tokyo.

Many Africans who make it across the Mediterranean end up in street camps in Italy or elsewhere in Europe. But far more either fail to get across the Mediterranean or do not even get that far. Twenty-eight-year-old Fatmata, from Freetown in Sierra Leone, fell into the hands of slave-traders who tortured her as she attempted to cross the Sahara. She eventually managed to escape, but she was recaptured and imprisoned in Algeria. She escaped a second time and, giving up on ever getting to Europe, pleaded with an NGO to help her return home. Two years after setting out, she was relieved to find herself back where she started. But her family disowned her, as she had failed to reach the promised land from where she might have sent them remittances. 'I was so happy to come back, but I wish I had not,' she lamented. Her brother said, 'You should not even have come home. You should just die where you went, because you didn't bring anything back home.'[14]

The families of those who either do not survive the trip or are sent back are invariably disappointed not to receive the remittances they desperately wanted. In many cases, they will have lost the money they invested in helping a family member reach what seems like a magical world of opulence and opportunity, from where money can be sent back and other members of the family can be sponsored. A Nigerian friend of mine who has long lived in London tells me that whenever she talks to her family back home, they attempt to foist on her a cousin or other distant relative who is desperate to start a new life in Britain.

This huge pressure from Africa is economic as well as demographic. The continent is experiencing rapid population growth, and as it gets richer, more people are in a position to scrape together enough money to try to send a family member to Europe. People living in urban areas tend to be more connected to the wider world and able to envisage life on a different continent, but increasingly, even those in remote rural areas are able to picture a different life. Connected to the internet through mobile phones, with social media like Facebook and communication tools such as Zoom and Skype, a totally different level of prosperity becomes visible and seems achievable to millions of Africans.[15] Vision leads to aspiration, which leads to action.

Beyond the demographic pressure, the economic attraction and the technologically assisted stimulation of aspiration, the other factor that drives migration from Africa is migration itself. Cousins who have already made the crossing successfully are encouraged or cajoled to help other family members to join them in their new homes. And once a community reaches a certain critical mass in the receiving country, it can smooth the way for others to come. This is done through working out the routes in, both legal and illegal, and helping with contacts, accommodation and information on arrival, all of which are crucial if a new immigrant is to find their feet.[16] An aunt might supply some of the funds for the journey. A cousin might provide a sofa for the first few nights. An old friend might help secure a job. Familiar shops, restaurants, newspapers and other goods and services allow the new migrant to feel comfortable in a social and cultural bubble that is redolent of the old country, which makes the transition easier and encourages more migration. What is true of Africans arriving in Europe today was true of Jews or Sicilians arriving in New York in the early twentieth century.

The scale of African migration to Europe also depends on European immigration policies. Between August 2017 and 2018, a total of 183,000 migrants reached Italy.[17] Many Africans arrive in Italy wishing to move on, yet a million or more remain. Italian populist parties had been on the rise for decades, but this explosion in immigrant numbers was the fillip they needed to take power in 2018. Uncomfortable though it makes some people feel to acknowledge it, Africa's population growth is already changing the ethnic and political map of Europe.

### *Within Africa*

For now, however, there is a much bigger flow of migration within Africa than out of it. Such movement has always existed. After all, national boundaries in the continent are mostly new and often correspond not to any human or geographic features but to arbitrary decisions taken by Europeans a century and a half ago. And as transport becomes cheaper and more accessible, more people than ever are moving between African states, with vast metropolises attracting people from both within and beyond national borders. In 1983, two million undocumented West Africans were expelled from Nigeria in the wake of economic strain caused by falling oil prices.[18] But in 2018, the best estimate indicated that there were half a million Ghanaians in Nigeria.[19] In South Africa, almost three million migrants reside. The largest foreign communities in Southern Africa come from elsewhere in Africa.[20]

Although this scale of movement would be inconceivable without the great population boom Africa has been experiencing, the immediate causes vary. In western and Southern Africa, migration tends to be driven by the search for economic opportunity, but the picture is more complicated. There may be well over a million migrants from Burkina Faso in wealthier Cote d'Ivoire, but there are also about half a million Ivoirians in Burkina Faso.

In East Africa, migration is more often caused by war: as of 2017, around 900,000 people had fled benighted South Sudan for Uganda, and about 300,000 to Sudan.[21] And the long-running civil war and instability in Somalia have resulted in a major outflow of people to neighbouring countries.

For now, Africans living abroad are more likely to live within their continent than beyond it. And for those contemplating a move, another country in the same region of Africa continues to be a more likely goal than Europe or America.[22] Much of this, no doubt, has to do with the relative ease of moving, say, from Uganda to Kenya, which is simpler, safer and cheaper than crossing the Mediterranean or getting the necessary visas and funds for a long-distance flight.

### *Africa's Future Fertility: The Great Unknown*

In the late 1980s and early 1990s, it seemed to some that with the end of the Cold War, we had reached the 'end of history'. Liberal democracy with a mixed economy was clearly the model that worked best, and the whole world would eventually end up in the same place. The most renowned exponent of this view was the political scientist Francis Fukuyama, who talked about 'getting to Denmark',[23] by which he meant that all of humanity would ultimately seek to replicate the prosperity, liberalism, political stability and human rights attained by the Danes. It is debatable whether the whole world, politically speaking, is converging on liberal, moderate Denmark, with its low levels of crime, its efficient economy, generous welfare state and stable democratic institutions. A much stronger case can be made, however, for a demographic convergence. From a population perspective, the world is moving towards Denmark, with lower infant mortality rates, higher life expectancy, a rising median age and smaller family sizes. In fact, some countries have overshot Denmark on these measures: Japanese live three years longer than Danes, and Greeks have nearly half a child fewer.

However, one exception is the fertility rate in sub-Saharan Africa. Although we can assume that the direction of travel will be Danish, the speed and extent of Africa's transition towards the low fertility rates common everywhere else is the great demographic unknown. Much about the future of our population hinges on this, including whether, by the end of the current century, there will be more than fifteen billion or just over seven billion of us. A 2014 study in *The Lancet* suggested that the global population would peak at just below ten billion in 2064 and fall below nine billion by 2100,[24] while the UN estimates that the global population will be nearly eleven billion, and still rising slowly, by the end of the current century.

Although we can be fairly sure that life expectancy will increase in Africa, the future of fertility rates is less certain. Today, Africa's women are still having an average of five children each, a figure that is little changed from seventy years ago and is unlike almost everywhere else in the world. According to the UN 'medium' forecast (which sits between higher and lower projections), by 2100, family size in Africa will shrink to replacement level: around 2.2 children. However, if that figure is above 2.5, Africa will have a population of five and a half billion. A 2020 study expects African fertility rates to fall below two children per woman by 2100, and for Africa's population to be around three billion by then.

Southern Africa is an exception. In South Africa itself, fertility is not much above replacement level, half what it was in the late 1970s. Family planning became part of government policy during Apartheid. The regime's birth-control programme was accused of being an effort by the white government to control the increase in the black population;[25] whatever its motivations, the programme was successful and South Africa's fertility rate began to diverge from the rest of the continent. Since then, the availability of family-planning services and their use has continued, and the fertility rate has continued to fall. While South Africa has many problems, it is not experiencing the explosive population growth that would otherwise have prevailed.

Although South Africa had a head start, the effect has spread to its neighbours. In Eswatini (formerly Swaziland), Lesotho and Namibia, fertility rates are close to three children per woman, while

Botswana is below this level. The latter country is a particularly notable case – it lacks the resources of South Africa, but is not far behind when it comes to promoting the use of contraception. It helps that the Botswanan health authorities have nearly fifty years of experience of working at a grassroots level. As ever where fertility rates have been lowered, the emphasis has been on educating women, although in recent times the focus has shifted to men's need to take responsibility. Trevor Oahile hosts a radio show about sexual and reproductive health that is targeted at boys. 'Boys must learn about menstruation at the same age as girls,' he insists.[26] Education continues to be the most effective way to help men and women control their own bodies and thereby their destinies.

However much governments, NGOs and activists may *do*, it is essential that they don't overlook what people *want*. It is true that a government could impose harsh restrictions like China's one-child policy. But for most of Africa, where government resources are scanty, control is often light and the pro-natal culture strong, such a scheme would be unimaginably draconian. Without Chinese-style coercion, it is the choices that people make once facilities for birth control are available that count. In Southern Africa there is a desire for smaller families, but elsewhere in sub-Saharan Africa, this is not the case. The family size of around five and a half children per woman in West and Central Africa in 2015 actually *undershot* what was desired by about half a child.[27]

We shouldn't read too much into surveys of how many children people want – the responses are often shaped by cultural nuances – but they might point towards the next big shift in African fertility rates, with lower fertility spreading from the south to the east. In East Africa, women have nearly one fewer child than their sisters in West Africa, and more than a child less than those in Central Africa. Although Uganda and Somalia continue to produce children in line with countries like Nigeria, Niger and Chad, the world's fertility leaders, major regional players like Kenya and Ethiopia are showing steady downward progress. In Kenya, innovative approaches are attempting to overcome taboos, such as an app that answers questions about contraception, providing more specific and reliable advice than has traditionally been available. '[Before] I would just Google,' one young woman told a reporter. 'Some questions are just too hard to ask some people.'[28]

In Nigeria, by contrast, progress towards lower fertility rates is painfully slow. It took Iran and China barely a decade to move from an average of six children per woman to three. Nigeria was at around six children per woman twenty years ago, and it is around five now. The reasons are a mix of pro-natalism (people want to have large families, regardless of facilities that help them do otherwise), poor delivery of government services and political instability in some parts of the country disrupting what services there are.

If Africans really are culturally pro-natal, there will be no end in sight to population growth, but why would that be? It is worth bearing in mind that the continent has long been underpopulated. In such an environment, with plentiful land but a shortage of people and high mortality rates, any culture or society *without* a fiercely pro-natal ethic would simply not have survived. This is in contrast to places like China, where a shortage of land created a culture in which, once a smaller family was possible, the option was eagerly taken up. It may just be that Africans, or at least some of them, really are more pro-natal than Asians or Europeans, but there is no reason why this should be perpetual. Cultures change, like everything else, and they adapt with the passing of the generations.

Two points are worth making in riposte to those who despair at the slowness with which Africa's fertility rate is falling. First, while it has been slower than China in the 1970s and 1980s and Iran in the 1980s and 1990s, it has at least in some countries been fairly rapid compared to places including the UK, where it took almost sixty years to fall from six children per woman to three. If we expect each region entering the demographic transition to proceed faster than the last, Africa will disappoint, but much of the continent is still transitioning more quickly than other places have. Second, despite the great progress Africa has made, we can only expect limited decline in fertility rates in a country like Nigeria, for example, where female literacy rates may even be declining.[29]

## *AIDS: Tragedy and Triumph*

Despite the increased availability and affordability of treatment, the spectre of AIDS continues to haunt Africa. However, anyone who thought that it would roll back the great human tide has been proven wrong.

The precedents are manifold. The second decade of the twentieth century was marked by the deadliest conflict in European history, with armies slugging it out on the Western Front for more than four years. On the other fronts, particularly in the Balkans and Russia, armies wreaked havoc with the latest deadly weapons. And to top it all, just as the First World War ended, the Spanish flu pandemic killed millions more. However, Europe's underlying population growth was so strong at this point, with falling general mortality and high fertility, that the continent's population was *higher* in 1920 than it had been in 1910.[30]

A more recent example, albeit on a smaller scale, is the bloody civil war that has gripped Syria over the past decade. A tally of half a million deaths is shocking, but it represents barely a year of Syria's natural population growth. It is mass emigration, not death, which has caused Syria's population to fall. All this is far away from the Black Death in the fourteenth century and the Thirty Years War in the seventeenth century, when national populations were cut by a third or more and took decades or even centuries to bounce back.

In today's conditions of improving food, water and healthcare, even the worst calamities are mere pinpricks against the power of demography.

Globally, it is estimated that thirty-five million people have died of AIDS, including about twenty-five million people in sub-Saharan Africa.[31] Over a period of nearly forty years, the disease caused something above 600,000 deaths a year in Africa. This represents around 2 per cent of the annual current number of births in the region, which explains why it has barely dented African population growth.

Suffering is, however, visible in the population data of countries in the south of the continent, where AIDS hit hardest. In Botswana between the late 1980s and the early part of the twenty-first century, life expectancy fell by a decade, despite other, positive developments in the country at the time. In Eswatini, the impact of AIDS was greater still, with national life expectancy cut by more than seventeen years, reversing decades of progress and taking the country back to where it had been in the 1950s.

The good news is that the global effort to develop treatments and make them available at an affordable price, along with education in sexual health, has reversed the situation, and life expectancy in Southern Africa has bounced back more rapidly than it fell. In demographic terms, the downward dip and recovery stands out against the steady lengthening of life expectancy across the developed world. To look at lines on a graph is valuable, but we should not forget the ruined lives the dip represents.

It is heartening that humanity has contained what once threatened to be a global pestilence on the scale of a biblical plague, even if it has not yet been fully defeated. In Eswatini, 27 per cent of those aged twenty-five to forty-nine are infected with HIV.[32] The United States deserves credit for the extraordinary work of the President's Emergency Plan for AIDS Relief, which offers anti-retroviral drugs to tens of thousands of people and has achieved a dramatic drop in the infection rate, but this would not have been possible without the collaboration of the country's authorities. 'The major, major breakthrough started coming when the current king understood it was a survival issue for his nation,' says Michel Sidibé, head of the Joint United Nations Programme on HIV/AIDS.[33]

In South Africa, the onward march of AIDS has also been reversed, following a disastrous period at the start of the twenty-first century in which the president, Thabo Mbeki, insisted that AIDS and HIV were not linked, resulting in treatment programmes being delayed. The data shows that life expectancy in South Africa has now risen above where it was in the early 1990s, its previous peak. And globally, deaths from AIDS have halved in the last fifteen or so years.

Ebola threatened to be another AIDS-style epidemic, though it spread much faster and was therefore potentially much more devastating. An outbreak was spotted in the West African country of Guinea at the end of 2013, and by the following summer it had spread to Liberia and Sierra Leone. There were isolated cases in other countries, including in Europe and North America, but these were contained – only in the three poor West African states did the disease seem to be out of control. The international community sprang into action, and all three countries were Ebola-free by early 2016. In total there were around 11,000 deaths, all of them highly regrettable, but the fast and effective response prevented this from becoming another Black Death.[34]

On the one hand, modern life clearly spreads disease, with travel more common even within poor countries, and international air travel taking infections to distant ports. On the other hand, modern conditions, and particularly the rapid response of the international community, support containment. Nobody can say for sure that some biological horror for which we are unprepared is not awaiting us, but humankind's impressive record gives grounds for optimism. For all the concern about Covid-19 and the disruption it has caused, the deaths it has led to have (at the time of writing) amounted to less than one in a thousand of the global population.

## *The Demographic Dividend: Population, Economics and the Future of Africa*

In early 2013 I was working on a project in Jakarta. The Indonesian capital is a famously polluted and crowded city and we allowed a good half-hour for the drive from our hotel to the office, but on one occasion, the journey seemed to take almost twice as long as normal. 'What's going on?' my colleague and I asked the driver. 'It's Valentine's Day,' he replied. 'Everyone is out on a date.' When we got back to our hotel and looked down from the thirtieth floor, the whole town seemed to be at a standstill.

Indonesia is overwhelmingly Muslim. Most Islamic authorities regard Valentine's Day as alien and encouraging of extra-marital relationships – it has no place in Islam or in traditional Javanese culture – yet the young people of this very young metropolis were embracing it, along with a range of Western practices. All across town, scantily clad girls were clinging to their boyfriends on the backs of motorbikes as they weaved in and out of the traffic. These girls will not have anything like as many children as their grandmothers did, and will probably have fewer than their mothers had, too.

Indonesia is enjoying what is known as a 'demographic dividend', which tends to occur when fertility rates start to drop. Because fertility rates were high until recently, many young people in their late teens and twenties are entering the workforce. However, unlike their parents, they are not rushing to have big families of their own. The fertility rate in Indonesia has fallen from five children per woman in the 1970s to just above two today. With a young, dynamic population, the workforce grows. Savings are often high, as young workers without many dependants start making provision for their pension. The dependency ratio of non-workers to workers falls, because these twenty-somethings are not having big families of their own, as their parents did. At this point, an economy has a real chance to boom; per capita income in Indonesia more than doubled in the first fifteen years of the current century.[35]

The demographic dividend is an opportunity but not a certainty. Indonesia has succeeded in grasping it, opening itself up to the global economy and attracting inward flows of capital. Its progress towards democracy has given investors confidence in its political stability. Other countries that have also reduced their fertility rate significantly have not taken advantage of similar demographic circumstances. Even before it was ripped apart by civil war, Syria was stymied by corruption and the ruling caste's refusal to change, and in this it has been typical of many countries in the Arab world. The instability in the region is, to some extent, a reflection of economic frustrations, but it also has deep demographic roots, which we will examine in a later chapter.

For Africa, the demographic dividend mostly lies in the future. Where families are still large, the young spend too much time and energy on their children to adopt the customs associated with economic growth. Women in poor countries with seven children may well want a washing machine or a fridge, but they are unlikely to have the means to acquire one, while any budget for education is stretched very thinly. But once family size declines, this changes. Better educated young parents limit the size of their family, in part to ensure that they can afford better schooling, nutrition and care for their offspring. Once Africa's demographic dividend becomes a possibility, everything will depend on whether the continent manages to adopt the Indonesian rather than the Syrian model. The

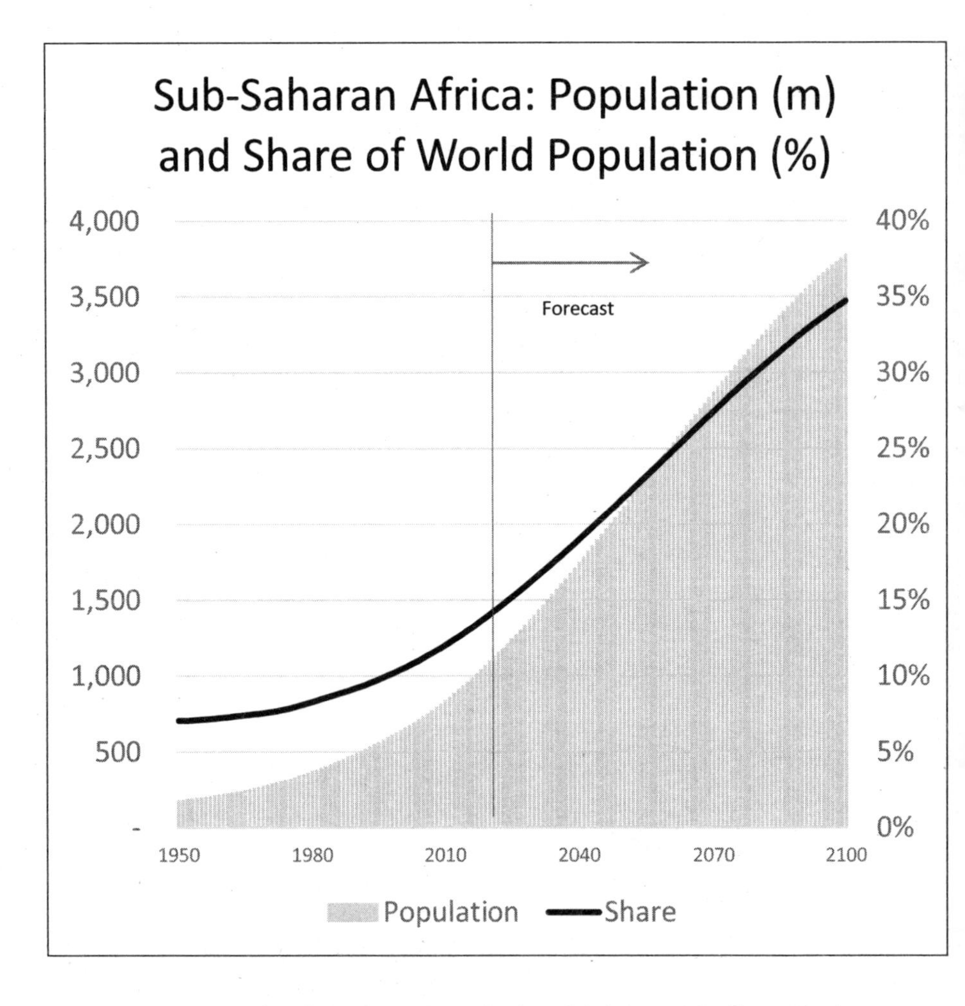


Source: United Nations Population Division, Medium Variant

The population transition is in its early stages in most of sub-Saharan Africa – explosive population growth is set to continue, and will only start to slow at the end of the twenty-first century. In 1950, fewer than 200 million people lived in the region; today there are more than a billion. The UN estimates that nearly four billion people will live there in 2100, but much depends on how quickly fertility rates fall.

Africa's rise as a share of global population has been stupendous, from around one in fourteen in 1950, to around one in seven today. By 2100, it is likely that more than one third of the world's population will be African.

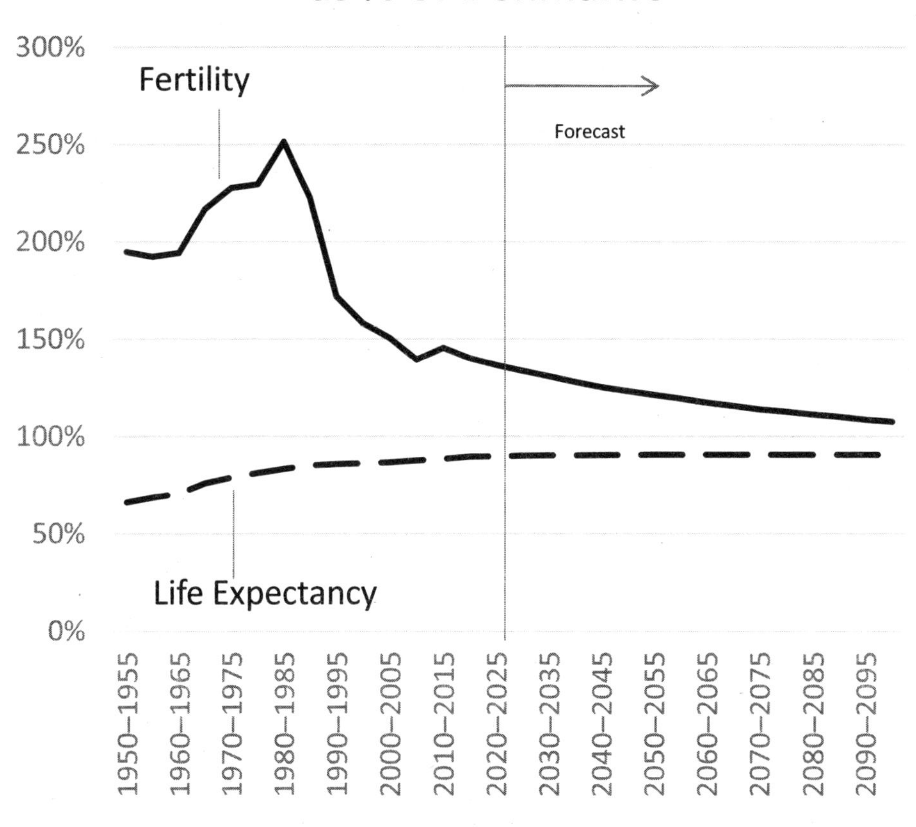


Source: UN Population Division, Medium Variant

The key demographic metrics tend to converge, with the biggest falls in fertility and the longest gains in life expectancy occurring in countries that have the largest families and shortest lives.

Demographically, the world is becoming more like wealthy, developed Denmark. In 1950, the global fertility rate was twice that of Denmark; today it is less than 50 per cent higher, and by the end of the current century, it is forecast to be barely 10 per cent higher.

The same is true for life expectancy. In 1950, the global average was two-thirds that of Denmark. Today it is only about 10 per cent shorter and the gap is set to narrow further, although probably more slowly.

potential is great, but for countries like Nigeria, a burgeoning workforce will not make a positive economic difference unless a more moderately sized generation follows it.[36]

### *Africa: Home of Tomorrow's People*

Whether Africa's population reaches 4 billion or 5.5 billion, this will be the greatest population event of our era. Britain pioneered the Industrial Revolution and the first modern population explosion in the nineteenth century, but the biggest events in human history have been the repetition of these two occurrences on a mammoth scale in China and Africa. China's industrialization and Africa's population expansion are reshaping the world in ways we are struggling to comprehend, and by 2100 Africa will look and feel profoundly different.

Their head start in the demographic transition gave Europeans a clear first-mover advantage, dynamism and population growth being the prerequisites for their subsequent period of global dominance. In many countries, such as the United States, Canada and Australia, Anglophone Europeans displaced indigenous peoples and created new societies in their own image. But the Africans will have *last-mover* advantage. The later a people move through the demographic transition, the faster mortality rates fall, and if this is not accompanied by a decline in fertility rates, population growth is astonishingly rapid.

Population expansion is Africa's greatest opportunity and its greatest challenge. The ability of its countries to absorb its emerging billions of people into productive work and integrate them into the global economy will define the fate of the world in the coming decades. The threat of over-population, resource depletion and desertification looms, but billions more well-fed and educated Africans could make an incalculable contribution to human progress.

The biggest challenges will be economic and political. If it is to end well – that is, if the majority of Africans are to lead long, healthy, peaceful and materially prosperous lives – it will require an extraordinary feat of human development. There could well be violent clashes over resources, mirroring similar European conflicts at the start of the twentieth century.[37] To succeed, African leaders will have to learn not only technological but also political lessons from the rest of the world. The African conflicts of recent decades, such as those in the Democratic Republic of Congo, have been the world's most lethal wars. Populations with many young men tend to be more volatile, and Africa will be young for a long time to come, making it prone to violence both *within* and *between* states. Fortunately, there are, these days, more productive ways to access resources and feed a growing population.

In 1950 there were more than two Japanese people for every Nigerian. Today there are almost one and a half Nigerians for every Japanese person, and by 2100, the UN expects Nigerians to outnumber Japanese people by *nine* to one. It's unthinkable that the world will not be radically different as a result, in terms of power, economics, culture and religion. *How* it will be different is a difficult question to answer. Africa is bound to have a much greater impact on world culture than hitherto. Perhaps Nigeria's Nollywood film industry will rival not only Bollywood but Hollywood itself. Great African writers may dominate the Nobel Prize for Literature. African universities and scientists may come to prominence on the global scene. The concerns of Africa, often marginal in past centuries, will become more central to the concerns of the world. And when a third of humanity is African, it will be difficult to deny an African country a permanent seat on the UN Security Council.

In Africa, the clash between Islam and Christianity is often acute and even violent, and these tensions will become more central as the continent grows in prominence. While conflict may remain at very local levels, if Africa experiences its own civilizational strife, it may be along religious lines, but there is nothing inevitable in this – Christians and Muslims have often lived peacefully side by side. Much of the tension is arising now because each represents a different version of modernity.

Whether the forthcoming decades are peaceful or violent, Africa's demographic expansion is assured. Even if African countries were able to reduce their fertility rates to a replacement level tomorrow, it would be decades before their population sizes stopped growing, thanks to demographic momentum – when population growth has been high, there are many young people to bear children, even if each has fewer. So the question determined by African fertility rates is not whether its population size will go on growing, but at what pace.

For now, the answer to this question is far from clear. As the African picture is so varied, the future is uncertain, but it is clear that most of the continent has a deeply embedded pro-natal culture. And this, perhaps even more than the region's slow economic development, explains why fertility has declined less in sub-Saharan Africa than anywhere else in the developing world. Furthermore, although African fertility rates are likely to keep falling, they may hover above replacement level for a long time. In the long term, the future will belong to cultures and societies that have a desire to bear and rear children. Europe, East Asia and much of the Americas seem to be failing this test; the great hope for humanity might depend on what Europeans once called the 'Dark Continent'.

One result of Africa's huge population growth has been something that tends to accompany population explosions: a rise in the number of people living in urban areas. Although cities like Lagos have mushroomed more than ten-fold since 1970, the continent as a whole has a long way to go before it reaches China's level of urbanization – and it is to this subject that we now turn.

# 3

# Urbanization

## *121: Chinese Cities With a Population of Over a Million People*[1]

On the eve of the First World War barely a dozen cities in the world had a population of over a million people.[2] Those that there were had grown exponentially as the populations of Europe, North America and Japan had expanded since the start of the nineteenth century. In 1914 any educated person would have heard of the world's most populous cities and been familiar with their locations and chief characteristics. Their names, New York, Tokyo, London and Paris, are still known around the world.

A little over a century later, 121 cities in China alone have a population of at least a million. At some point in the 2020s, a further one hundred Chinese cities will pass the same mark,[3] but even the most knowledgeable people could be forgiven for barely having heard of the vast majority of them. Indeed, when there are so many megalopolises, and when the dimensions of the biggest have grown so dramatically, the very definition would benefit from revision.

Most of the biggest cities in the world have grown from rather modest conurbations, but some have been founded from scratch. Brasília, the capital of Brazil, was created in 1960 and now has a population of around 2.5 million. Abuja, the Nigerian capital, barely existed before 1980 and already has a population of a similar size. And India has around forty cities with a million or more inhabitants, many of which are barely known to the wider world.[4] The really huge city, once a famous rarity, is now commonplace.

What counts as a city is not exactly defined and neither is the measurement of a city's population. First, the boundaries of a particular city may not be all that clear. In the case of London, for example, should we count its population as those who live within the M25, within the boroughs of Greater London or within commuting distance of the West End or the City of London? Second, it is not obvious how a village should be distinguished from a town, or a town from a city. And third, the data on the population of conurbations is better in some parts of the world than in others. Many cities have relatively small populations but significant workforces who commute in. The city of Luxembourg has a population of around 125,000, but a much larger number of people commute in every day from Belgium, France and Germany, not to mention those who travel from other parts of the Grand Duchy itself.[5]

The urban theorist V. Gordon Childe argued that ten characteristics marked out cities when they first emerged: extensive and dense populations, craft and artisanal specialization, capital intensiveness, large-scale buildings, a socio-economic class exempt from manual labour, record-keeping and knowledge creation, writing, artists, trade, and security on the basis of residency rather than kinship.[6] This is less helpful in today's world, where entire countries have many if not all of these features. Perhaps the best we can say of defining a city is that we know one when we see one.

### *China's Anonymous Megacities*

Nanchang, in south-east China, is bounded by the Jiuling Mountains and Poyang Lake, the biggest body of fresh water in China. The city played a prominent role in the rise of communism, staging a notable rebellion in 1927,[7] for which it earned the nickname 'Heroes'

City'. As well as holding a special place in the hearts of China's communist establishment, it is also an important site for Taoists.

So, clearly, Nanchang is unlike Brasília or Abuja, recently thrown up from next to nothing; it has a long and distinguished place in Chinese history. But only in recent years has it reached anything like its current size, with around five million people living in its wider urban area. As recently as 1970, the city had well under a million inhabitants, while the old city on the eastern side of the Gan River has recently been completely overshadowed by a new one on the western bank. It is rather like what happened when London outgrew 'the City', first spreading towards and engulfing the City of Westminster and then eating away at the surrounding county of Middlesex before crossing the River Thames to expand southward. But in the case of Nanchang, this has happened over a few decades, and without most of the rest of the world either knowing or caring.[8] It is perhaps rather Eurocentric to think of Nanchang as anonymous just because it is known to so few Europeans, but as the number of Chinese cities with a million people surges towards 200, we can expect many of them to remain little-known even to those who live within China's borders.

Like many other Chinese cities, modern Nanchang is the product of the exodus from the countryside to town, but this has hardly been a uniquely Chinese phenomenon. In 1800, just 6 per cent of the world's population lived in towns. By 2007, half of humanity was urban, and estimates suggest that it will be more like two thirds by 2050.[9] The Chinese countryside experienced a great population surge in the mid-twentieth century, and although the fertility rate slowed from 1970 – particularly following the introduction of the one-child policy from 1979 – it took time for the cohort of rural youth to reduce in number. Even when the countryside was no longer booming, village dwellers continued to move to towns in search of opportunity.

On an even greater scale than Nanchang – and not much better known in the West – is Chongqing. Its population of nearly sixteen million people makes it about twice as large as Greater London or metropolitan New York.[10] The city dates back to the early days of Chinese civilization, but its population was a quarter of its current size as recently as 1990. Like other Chinese cities, it has grown as a result of the general rise in population and the movement of rural people to the towns. Its growth has also benefited from the state-directed effort to redirect economic growth away from the booming coast and towards the Chinese interior. As well as indigenous businesses, Chongqing has managed to attract the likes of Ford and Microsoft, creating jobs with salaries that bring peasants and small farmers to the city.

Economic opportunity has always drawn people from the countryside to the town, but this pull has been amplified by the push of demography. As infant mortality rates plunge and rural populations surge, the division of land into ever smaller plots becomes untenable. The countryside is unable to absorb its population, and the great urban drift begins.

In the mid-1970s, 80 per cent of China's population lived on the land; now, over 60 per cent is urban.[11] This may have happened faster and on a greater scale in China than elsewhere, but the effects have been similar all over the world, and urban human beings are simply different from rural ones.

### *The Rise of Homo Urbanus*

When a society changes from being predominantly rural to being predominantly urban, the shift is not just in where people live – it involves a change to its mindset, to its very soul.

Some years ago, I was a governor at a primary school in an area of London with a large Bangladeshi community. One of the teachers complained that when a trip to the countryside was planned, many of the Bangladeshi parents would withdraw their children for the day. I discussed this with several of the parents and got the impression that, having come from a poor and relatively remote province of Bangladesh, the parents could see no sense in visiting a rural area – they prided themselves on having left the countryside behind.

I was reminded of this when, some years later, I was working in Serbia and a British colleague asked a local manager in the national telecoms company if he ever considered leaving Belgrade and returning to his rural roots. My colleague, like me, was the product of an urban childhood, while our Serbian friend, who had a relatively prestigious head-office job and a house in a middle-class suburb, was just one generation away from life in a remote village. He looked at my colleague as if he was mad – a return to his rural roots was the last thing he aspired to. Along similar lines, a friend told me about his father-in-law, who had decided to retire to Jamaica, from where he had come to the UK decades earlier. When one of his children suggested that he might buy a smallholding and keep chickens, a suburban pastime of two families I know in North London, he was shocked. 'I don't want to go back to where I came from,' he replied.

For those people who were brought up in a town and whose parents and grandparents were likewise urban, the countryside becomes somewhere to escape *to*, not *from*. And those of us who were raised in societies that have long been predominantly urban have forgotten the extent to which escaping to the countryside is a First World luxury. We think of rural life as a romantic idyll rather than a hard grind in all weathers to extract a basic subsistence from the land. But nowadays, the town–country divide that marked earlier eras is gone. In the UK, for example, people living in the countryside are rarely far from a city, and any distance is trivial if you have access to a car.

Furthermore, rural areas today in the UK often have the same educational, utility and healthcare services as the cities.

Back before radio, the railway, the motor car and decent roads, people living even in the heartland of south-east England were cut off from much that modern life has to offer. Even in the 1940s, when my mother was evacuated from London during the Blitz to live in Bedfordshire, which is now an easy commuting distance from London, there was no electricity in her home and most villagers had to pump their water from a well on the village green. The lack of a sewage system meant my grandfather had to bury the family's waste in the garden. Although that was all about to change, little more than a third of British homes had telephones as recently as 1970.[12]

Benefits that were once the preserve of the city are now available in rural areas, a trend enhanced by the arrival of the internet. Online magazines and podcasts are as accessible to those living outside cities, subject to local bandwidth; providing the internet is working, the experience of rural life need no longer be much different from urban life when it comes to entertainment or stimulation. An e-reader makes up for the lack of a local bookshop, Netflix for the absence of an easily reachable cinema.

Differences in political and religious attitudes used to separate the city from the countryside, with people in rural areas more bound to tradition. The old German saying 'Stadtluft macht frei' means 'City air is liberating'. Countryfolk may have felt themselves 'free' despite their more conservative attitudes, but in earlier times much of rural Europe was gripped by feudalism, reflected by deference towards certain people and resistance to change. Even the way we look at the world today, seeing it as having discernible patterns that can be grasped through physical laws and mathematical models rather than the intervention of deities and demons, was the creation of the early modern city.[13] At the very least, people in cities were more literate than those outside them. For example, in the 1860s, when almost everyone in Berlin could read, a third of those in West Prussia could not.[14]

Urban and rural living have converged in the developed world, but this is less true elsewhere. For people in the developing world today, leaving the countryside for the city, whether in Asia, Africa or Latin America, still holds the promise of opportunity, just as it once did in Europe. Moving to the town increases the chance of paid employment, which allows a better standard of living than subsistence agriculture. It also increases the chances of your children gaining a decent education that will let them find work that requires brains rather than brawn. 'My dad and grandpa spent their lives crouching in the rice paddies,' a Jakartan taxi driver once told me. 'Driving a cab is much better than that, and now my son works in an air-conditioned office.' Those of us who struggle to understand the joy of sitting at a desk in a temperature-controlled environment, or even behind a wheel, haven't spent years labouring under the baking Javanese sun.

This perspective is captured by one of the unfortunate people caught up in Mao's murderous Cultural Revolution of the 1960s and 1970s and forced to make the reverse journey, from a modest but relatively comfortable life in Beijing to a life of peasant subsistence in Inner Mongolia:

> There was no shade to shelter from the blazing sun . . . We would get up at about four in the morning . . . swarms of mosquitoes would attack most relentlessly. There was no escape from their attacks in the sweltering heat. While we could not feed ourselves by farming, we helped contribute to the food supply for the Gobi insects with our own bodies.[15]

Karl Marx's disparaging references to the 'idiocy of rural life' may sound condescending to modern ears but it is true that the peasantry were not only less literate and less educated than the town dwellers, they were also less inclined and less well placed to make revolution. Peasants, when roused, had *revolts*; it was urban dwellers who had *revolutions*. Revolts tended to be localized and poorly organized, and they were invariably put down by the government. Wat Tyler's Peasants' Revolt in England in 1381 and Yemelyan Pugachev's insurrection in Russia in 1781 are typical. Tyler ended up with his head displayed on a pole on London Bridge, while Pugachev was beheaded, drawn and quartered in a public square in Moscow. Their revolts were expressions of anger rather than of political intent, and they were swiftly crushed by the authorities.

Small wonder, then, that Marx saw the revolutionary future not in the hands of those he viewed as ignorant, reactionary peasants but as driven by a new class: the proletariat. This new class of urban, industrial workers consisted of people who had a basic level of education. According to Marx, this meant they would be more conscious of their class interests and more able to act on them. The urban proletariat was therefore in a position both to organize *en masse* and to take control of the all-important towns.

The industrial working class may not have turned out to be the heralds of revolution, but the great revolutions have always happened in cities, most famously in Paris, St Petersburg and Tehran. And when a counter-revolution came, as in France in 1848, it tended to arrive from the countryside, in the form of an army of peasant soldiers led by feudal aristocrats.

Although towns are overwhelmingly associated with the future and the countryside with the past, it is worth noting that precisely *because* they are repositories of civilization, the towns are also the places where the past is preserved, in the form of monuments, libraries and museums.

Towns are not just associated with political and economic change. They have, since the earliest times, been associated with civilization, the very term 'civilization' deriving from 'civitas', the Latin word for town. People create and innovate best when they are able to rub shoulders, converse and exchange ideas, as is the case in towns. In the countryside, interacting with people beyond their immediate area requires them to use their scarce resources. The future was invented as much in the coffee houses of late seventeenth-century London and the cafes of early twentieth-century Vienna as in university laboratories. The sinews of the state, such as writing and taxation, along with the arts and the sciences, all developed in the city. People in cities were better able to defend themselves than those in isolated villages, hamlets or farmsteads. The city effectively creates an economy of scale in terms of security, which meant that in times of lawlessness, capital was more likely to be preserved in the city than in the countryside. The city also benefits from economies of scale in utilities such as water and sewage, healthcare and education.

The extent to which cities flourish is, therefore, a measure of a civilization's sophistication. Each city needs to import vast quantities of building material, food and water. Conversely, when a civilization is in decline, it has neither the requirement for great cities nor the means to sustain them. The great example of this in the West was the decline of cities during the fall of the Roman Empire. Rome was a striking example, although many other cities in the empire shared its fate. Having had a population in the first century CE that probably exceeded one million, Rome had barely thirty thousand inhabitants three centuries later.[16] London probably lost two-thirds of its population in the decades after the number of inhabitants peaked as a Roman city in around 150 CE.[17] Nothing more starkly demonstrates the decline of the Roman Empire than its demographic decay. By 1200, Paris, the biggest city in the Christian West, had just over a hundred thousand people, barely a tenth of Rome's population at its peak. The technology of Western Europe at that point could not support anything larger.

Modern humans are so overwhelmingly the product of urban culture that we hardly see how striking this is. We forget that while the city is thousands of years old, it has only recently housed any more than a tiny share of the population. Just 1.6 per cent of Europe's population was urban in 1600, and a little over 2 per cent in 1800. But by 1801, 10 per cent of England and Wales was urban, and the majority of their populations lived in cities well before 1900.[18] In urbanization, as in the rest of the demographic transition, the British Isles were pioneers;[19] a majority of the Earth's population has only been urban since the early years of this current century.[20]

### *Urbanization and the Environment*

As we have seen, naive urbanites think of the countryside as perennially rosy and associate it with leisure pursuits rather than hard graft in the fields. They also tend to think of rural life as somehow 'natural', and that life in towns is detrimental to the environment. In fact, nothing could be further from the truth.

It is certainly the case that pre-modern societies have low carbon emissions per capita. Those who rarely travel other than on foot or by bicycle, who have never had access to electricity and who would not know what to do with a washing machine if confronted by one are light in their use of the planet's resources. But when modernity arrives, its advantages are far more efficiently and sustainably enjoyed by those who live in the cities.

The average American town-dwelling family, for example, will drive fewer miles and probably own fewer vehicles than its countryside-dwelling cousins, who may well need to get in the car each time they want a loaf of bread or a pint of milk. Urbanites are more likely to use public transport for most of their daily travel needs. Their journey to school or to work is probably shorter and less reliant on the combustion engine. The share of Londoners who travel to work by car, for example, is much less than half that of every other UK region.[21] Services such as running water and sewage systems, and the provision of electricity and mail, can be more effectively delivered to town dwellers, and require fewer miles of pipe or cabling and less tarmacked road.

Town dwellers are likely to live in smaller, better-insulated homes than those in the countryside, so they will be more fuel efficient and their overall emissions lower. A 2004 study, for example, found that Londoners' carbon emissions were around half the UK national average,[22] while New Yorkers are reckoned to emit around 30 per cent less carbon than the average American.[23] Highly urbanized New York has the lowest per capita emissions of any US state.[24] As the urban designer and planner Peter Calthorpe put it, 'The city is the most environmentally benign form of human settlement. Each city-dweller consumes less land, less energy, less water and produces less pollution than his counterpart in settlements of lower density.'[25] Our tendency to live in towns and cities allows half of all humans to live on less than 3 per cent of the world's surface.[26]

Modern cities are, of course, still huge consumers of energy; [27] it's just that once people start leading modern lives, those in cities do so most energy efficiently. The real gas-guzzlers are not the rural peasants in poor countries but the country dwellers in the developed world, who tend to drive great distances for life's necessities and pleasures while living in less compact and well-insulated homes.

Urbanization also presents more immediate advantages for nature: the emptying of the countryside provides a respite. In the Pyrénées-Orientales in southern France, close to the Spanish border, there are acres of wild hillside where the only signs of previous intensive cultivation are the slowly collapsing terraces that were once tiny agricultural plots. Elderly people in the area can still remember when these long-abandoned hillsides were carefully tended. The villages in this region are not all emptying out, but its inhabitants are leaving agriculture, and many of them have upped sticks and left the valley to head for the nearest big town or beyond.

In place of crops, natural flora and fauna have returned, with wolves and bears spotted in the area for the first time in decades. The bears have been reintroduced to the Pyrenees as an experiment, while the wolves seem to have made their own way from northern Italy, taking advantage of increasingly empty tracts across southern France.[28] Wild boar also proliferate, and their numbers are only kept down by the local enthusiasm for hunting. Nature bounces back when humans leave the countryside and head for the town. A third of land that was once farmed in the Soviet Union has now been abandoned to nature; New England, deforested by early European settlers, saw tree coverage rise again in the twentieth century, from 30 per cent to 80 per cent of land coverage.[29]

### *A Reversible Trend?*

It is possible that urbanization could go into reverse. Two things could cause this. First, we could have some kind of civilizational collapse, whether caused by a pandemic, a financial implosion or by one of a number of disasters that could strike humanity at any time. In such an event, our extraordinarily complex civilization would probably break down. Supplies of food and water and everything else that makes urban life possible would dry up, and those who survived would find themselves trying to subsist on the countryside's natural resources. The Covid-19 crisis has only had a moderate impact on human numbers, but it still led to the temporary desertion of some city centres.

It is too early to say whether Covid-19 will result in long-term urban decline, but such a decline has happened before in times of crisis. Apart from the collapse of the Roman Empire, another example is the chaos of revolution and civil war that beset Russia in the early twentieth century. Recently proletarianized peasants returned to their villages – it was only with the industrialization of the Stalin era that the process of urbanization resumed.

A global pandemic on a greater scale than that which began in 2020 could cause people to flee the towns for the countryside. This would be a modern version of the movement, by those who could afford it, from London to Hampstead in times of plague over six hundred years ago. If such an urban exodus occurred, once the pandemic had passed and the population had been reduced, towns might be unattractive, violent places; rural life, by contrast, might be more viable. The fact that such thoughts are the stuff of 'survivor' fantasy fiction and of 'prepping' in case of a breakdown in supplies and law and order does not mean it could never happen.

Although the occurrence of such a global disaster may seem improbable, there is historical precedent for such civilizational collapse accompanied by urban breakdown. Indeed, once thriving but long deserted cities are the mainstay of archaeology and cultural tourism, from the pagodas of Siem Reap in Cambodia to the Mayan ruins in Mexico's Yucatan Peninsula. The failure of these and similar urban centres was either the result of ecological factors or of political issues. In a world where people on different continents barely knew of each other's existence, these were local events. In an interconnected world, the ecological and political calamities of a similar collapse would play out on a global scale. We enjoy the great benefits of globalization, with near-instantaneous financial trades occurring across thousands of miles and millions of people airborne every day, but this means that problems, whether biological or economic, can spread faster and further. We are more aware of this since the arrival of Covid-19.

The second cause of a reverse in urbanization might be a forwards rather than a backwards step. Technological progress might either mean that people could commute over longer distances with great ease or that they do not need to do so at all. With modern teleconferencing and IT, people can increasingly work together without being in the same place, and the development of hologram technology might make this even more of a reality. People might choose to live further from their workplace, either because of a preference for rural life, because accommodation is more affordable or to avoid a troublesome and costly commute. The Covid-19 crisis has accelerated this change.

Nevertheless, people might choose to continue living in or close to towns even when they no longer need to do so. The technology that enables remote working is already quite advanced, but workers still commute in their millions every day from the suburbs into city centres to benefit from face-to-face interaction with colleagues, customers and suppliers. It may be that these gains cannot be eliminated by technological innovations. Likewise, business travellers still fly to meetings in distant cities that could be held by teleconference. Physical proximity remains crucial when it comes to building and sustaining interpersonal relations in the business world, never mind the social one. Despite this, a move away from city living and commuting could be seen even before the coronavirus pandemic. There was a notable decline in the use of London's Underground system, with its finances under increased pressure.[30]

The yearning for a simple, quiet life, with more living space and access to the outdoors, might seem like a rich person's indulgence, but a weariness with urban existence seems to resonate with many. Take this lyrical ode to country living by someone who left London for the Welsh borders:

> My nerves shredded by the city, I'd handed back the keys to my tiny flat near Regent's Park and found a converted granary to rent, just outside the village of Cwmyoy. It stands at the top of a steep track halfway up a mountain, its elevation giving the feeling of

> being on the prow of a ship in a rolling, viridescent sea. Most mornings, before writing, I'd set off for a five-mile walk. This was initially a heads-down affair – with headphones on – akin to going to the gym, exercise as necessity, to get out of the way. But something shifted. The walk quickly became my life. The headphones were discarded . . . And the longer I am here, the less I can imagine being anywhere else. 'It's like you are marooned in paradise,' said a friend on the phone. And, in so many ways, he was right.[31]

A decline in the number of people living in major cities would not be unprecedented, and we needn't go as far back as the fall of the Roman Empire or the Black Death to find examples. Between 1939 and 1991, the population of London fell from eight and a half to six and a half million, before the trend reversed and numbers rose again.[32] In London's case, this return to population growth was caused by a wave of immigration from overseas that, by the 1990s, more than offset the outflow of people of British origin. There was a similar trend in Amsterdam, where a post-war exodus to improved housing beyond the city boundaries until the 1980s was followed by an inflow of North African and Turkish migrants and then, from the late 1990s, an inflow of young aspiring professionals.

Today, the influx of people into cities has caused property prices to rise, which encourages more people to leave. Many of the young professionals who flocked to the cities ten or fifteen years ago choose to move away once they have families, in search of more spacious and affordable accommodation in family-friendly areas. Although they are followed by a new cohort of young professionals drawn to the excitement of urban life, the later cohort tends to be smaller than the preceding one, given the recent overall decline in birth rates in most developed countries.[33] These contradictory and offsetting trends could lead to a decline in the size of cities in countries like the Netherlands and the UK, although when people leave cities like Amsterdam or London, they are likely to settle within striking distance of their old haunts.

However much cities in the developed world may rise and fall, urbanization will continue in much of the developing world. In fact, this trend is happening on a vastly greater scale than any potential flight from European and North American cities. For every aspiring writer leaving London for the Welsh Marches, there are hundreds of Nigerians eager to get to Lagos – and to London. So from a global perspective, humans will increasingly be town-dwellers, with urbanites probably making up three-quarters of the global population by the middle of the current century.[34] For the foreseeable future, any meaningful reverse in that trend will only be among the world's most privileged.

In sub-Saharan Africa, for example, around 60 per cent of the population is still rural, way down from 85 per cent in 1960 and moving in one direction.[35] Huge megacities like Lagos have mushroomed in recent decades and now boast far bigger populations than have ever been seen in Europe. Lagos is believed to be home to fifteen or twenty million people, depending on where its borders are drawn.[36]

### *Cities of the Future*

The people of the future may become increasingly town-dwelling, but the towns where they live will not remain the same. For a start, cities in the developing world will likely become more like those in Europe, North America and the richer parts of Asia. As the incomes of urban dwellers in the poorer parts of the world increase, their consumption patterns and demands on public services will rise. Just as the slums of European and North American cities have been cleared, we can expect the same thing to happen in Kinshasa and Jakarta. This is not to say that there will be no urban poor, but basic conditions will improve. Running water, adequate plumbing and indoor toilet facilities will become the norm rather than the exception. Public transport will expand, reducing the frustration and pollution of traffic jams. Motor emission standards will rise, which will further reduce pollution. City dwellers will demand these improvements, and their municipalities and governments will increasingly be able to supply them.

In richer countries, the provision of healthcare, education and transport possibilities is making the countryside more like the cities, while towns are becoming more like the countryside. When the London Underground network was extended in the early twentieth century, people who lived and worked in the inner city were able to move out to the suburbs. There, they could enjoy their own gardens and *rus in urbe* – 'the countryside in the town', an allusion that went back to Roman times. During the 2020 coronavirus lockdown in London, living in a house built just before the First World War and with sixty feet of garden, I was grateful to that development, as were millions of my fellow Londoners.

The arrival of the motor car in the United States and Europe allowed the development of outer suburbs, commutable for city workers but with even more open spaces and better amenities. These may be criticized for the creation of 'urban sprawl', but they offer a high standard of living for millions. And as cities become wealthier, space is allocated for the development of urban parklands, for urban wildlife and even for urban agriculture.[37]

The appearance of the cities of the future has always been the subject of much speculation, because cities are inherently futuristic. We can be reasonably certain that there will be more of them and they will be bigger. Whatever happens to fertility rates in the coming decades, the world's population is, in the absence of a global calamity, destined to keep growing at least into the second half of the current century. Most of these extra people will end up living in cities, as will many who are currently on the land. There may also be a consolidation of cities; some, like Manchester and Toulouse, have thrived in the post-industrial world, while others, like Detroit and Middlesbrough, have struggled.

As cities grow, they tend to erode the surrounding agricultural land, a change that will be increasingly resisted if food becomes rarer. But this is still a long way off, as the enormous rise in land values when fields are reallocated for urban use illustrates. There are examples in the UK of the same fields being worth about 150 times more once they are deemed usable for housing.[38]

Cities have always been net importers of food and energy, engaging and absorbing the resources of their hinterlands. Although the recent trend towards urban food production may just be a short-term fad, there are reasons to think it might not be. Converted warehouses fitted with artificial lights are used to grow food on multiple layers and without the herbicides and pesticides required in conventional agriculture, and with efficient use of water and nutrients. Another great advantage of urban food production is that its output can be sold without the financial and environmental expense of long-distance distribution. Similarly, just as the transportation of coal by barge from Newcastle to London is a thing of the past, gas may soon no longer be piped to cities or converted into electricity and transmitted. Solar panels are already installed on roofs and might one day be added to walls. Along with the possibility of privately installed mini-windmills, such innovation might allow the city to produce much of its own energy.

If electric cars become the norm, cities will be much less polluted. This trend is already long-established in developed countries, thanks to the replacement of coal fires with more environmentally friendly forms of heating. Another seemingly fantastical possibility for urban transport, flying cars would certainly ease ground congestion. More immediately, urban transport is already being transformed by car-sharing services and taxi services that are enabled by the mobile phone and internet revolution.

Car ownership could become less common. Self-driving cars will further reduce the number of vehicles on the road and liberate valuable urban space that is currently devoted to parking. Many cities are seeing the rise of cycle lanes, along with the availability of rented bikes and the emergence (or re-emergence) of trams, all things that reduce the need to own a car while also air pollution. Walking past suburban front drives that have been concreted over to hold as many cars as possible, I find myself wondering what they will look like in thirty years' time. Will they return to what was once known as a 'front garden'? This would add to the natural content of cities and help cool them, something that will be particularly valuable if the planet continues to heat up.

It is important that we learn from the mistakes of the past. Future cities need not be chaotic and unplanned like Dickensian London or modern Jakarta or Lagos, but neither need they be soulless and socially alienating like the housing developments that accompanied post-war slum clearance in Europe. More imaginative balances between spontaneity and planning will be struck, and the cohabitation of humans and nature will be an important part of that balance.

Successful cities often define their nation states while struggling to fit within them. London dominates the UK, although it is increasingly diverging from its hinterland economically and politically, and the same can be said of Paris. Lagos has an outsized impact on Nigeria, just as Mexico City has on Mexico. The great conurbations are usually ethnically diverse and young, attracting both the nation's brightest and best and people from further afield. But they are often looked at suspiciously and even resented by those who live elsewhere.

The UK referendum to leave the EU in 2016 provides a good

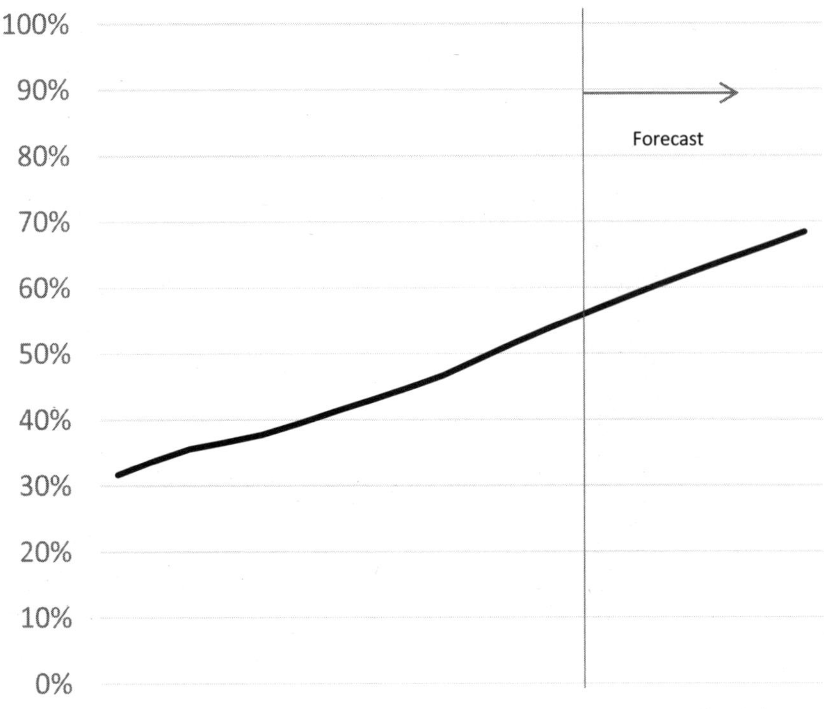


Source: United Nations Population Division

Since the development of agriculture, most people have lived on the land, with towns containing a small share of the total population. However, by the middle of the twentieth century, a third of the world's population lived in urban areas, and we have recently passed the point at which more than half the people on the planet were town-dwellers. The trend will continue, and by the middle of the current century around 70 per cent of people will live in towns.

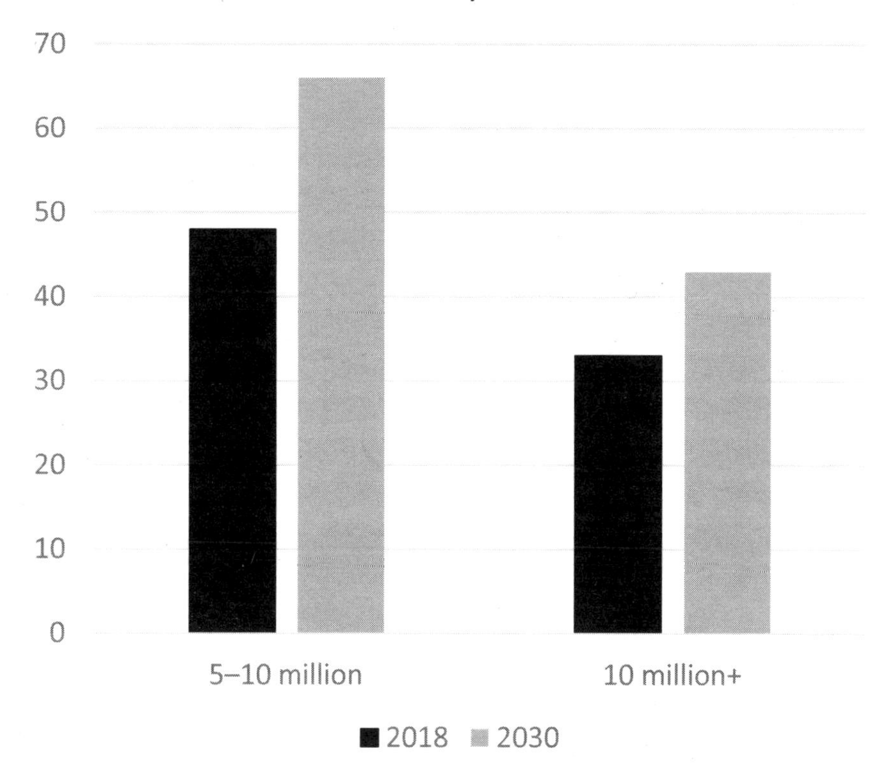


Source: United Nations World Cities

The world's cities are getting both bigger and more numerous. There were forty-eight with populations of between five and ten million people in 2018; in 2060 there will be sixty-six. Over the same period, the number of cities with populations above ten million people is expected to grow from thirty-three to forty-three.

example. London was strongly against 'Brexit', while many other parts of England were strongly in favour. The British prime minister Boris Johnson has described it as a vote not just against Brussels but against London, despite his representing a Greater London constituency and having previously been London mayor.[39] Resentment at the perceived pretensions of the big cities are as old as the cities themselves; they are accused of acting like sponges and sucking up the human and material resources of the rest of the country. In response, they point out that their productivity and economic success account for a disproportionate share of the national budget, supporting the educational, welfare and health services of the whole nation.

### *Urbanization and Demography*

Cities and population have a two-way impact on each other. Cities tend to grow when rural populations expand and overflow into towns, but once people have moved to towns, their lifestyles and family sizes change, which feeds back into population change.

So the rise and fall of cities is shaped by big demographic trends, but it also shapes them. The movement of a burgeoning rural population into cities happens at the start of the first demographic transition, when fertility rates are high and mortality rates are plummeting, as in Britain in the early nineteenth century and modern Nigeria. The city absorbs much of the growing population that cannot be accommodated on the land, but it then changes the behaviour of those who move to live there.

The relationship between urbanization and mortality is more complicated. At one time, a city's unhealthy air and open sewers made it a sink of disease and elevated its death rate. Cities attracted people and used them up, just as they did other resources. This was true, for example, of London in the eighteenth century, which required continuous inward flows of people to maintain its size. In the middle of the nineteenth century, life expectancy was just twenty-six in Manchester and Liverpool and thirty-six in London, but in England and Wales as a whole it was forty-one.[40] Around 80 per cent of Moscow's population is estimated to have died of the plague in 1654.[41] In the following year, the disease killed 20 per cent of Londoners, but only 13 per cent of the English population.[42] As late as the first decade of the twentieth century, death rates in cities were a third higher than those of the countryside.[43] The move to the city often meant a worse living environment, a poorer diet and an earlier death, yet cities could absorb great numbers, and so they were able to become more numerous.

Over the course of the twentieth century, the opposite became true: urban lifespans increased as cities became cleaner and healthier. Where once they were breeding grounds for disease and plague, they are now centres of accessible medical care and education that tends to enhance longevity.

Eventually, as countries develop and levels of healthcare and education improve in rural areas as in urban ones, the difference in life expectancy narrows again, and the impact of pollution and the stress of urban life can even make village life seem healthier. On the other hand, where town-dwellers are richer, this contributes to their living longer. The inhabitants of London and south-east England enjoy longer lives than those in any other region of the UK, despite the stresses of urban life and the poor air quality that many have to suffer.[44]

While the impact of urbanization on mortality is complex, the impact on fertility is less ambiguous. For a rural farmer, an extra pair of hands is welcome, whereas for urban parents, another child reduces the resources that can be invested in each member of the next generation. In much of the world, town dwellers are better educated and their children are less likely to die at an early age, and they tend to have fewer of them for both these reasons. People who do not expect to lose any of their offspring and whose education provides them with life opportunities have smaller families. Furthermore, family planning programmes are more easily administered in urban areas than in remote rural districts where people are difficult to reach and may be subject to societal conservatism and patriarchal pressure. But again, once a country is fully developed, the differences between town and countryside diminish.

There are cities in the developing world where fertility rates are surprisingly low. In Kolkata, the Indian megacity, the average woman has just 1.2 children, around half the national average.[45] Women there express opinions about the drawbacks and inconvenience of having a second child that were, until recently, regularly heard in Paris or New York, but rarely in a city in the developing world. In other Indian cities, such as Chennai and Mumbai, the fertility rate is not much higher. It's extraordinary to think that women in these still-poor cities, where millions of people live in slums, have fertility rates that are significantly lower than some of the world's wealthiest countries. However, even these places have not yet reached the fertility trough of Singapore.

# 4

# Fertility

## *1: Singapore's Total Fertility Rate*

'Why bother having kids when a child-free life is much easier?' So writes the editor of a Singaporean magazine. 'All those sleepless nights, dirty mornings, shit and piss everywhere . . . not to mention a harder time focusing on your career.'[1] A spectre is haunting the developed world, along with parts of the developing world. But rather than too many deaths, it concerns too few births. The worry is that eventually the human race will be at risk of disappearing altogether. After decades of worrying about population growth, people are now starting to worry about the opposite.[2]

In some cases, family size may be something over which individuals have little control; they might want children but not be able to have them. Sperm counts appear to have fallen in developed countries by 50 to 60 per cent since the mid-1970s.[3] But 84 per cent of couples in the UK will conceive within a year if they have regular sex without contraception,[4] and this is despite the fact that this will, in many cases, be happening well after the woman's peak fertility. So a lack of children, in many cases, reflects an attitude expressed in the quote at the start of this chapter. And Singapore, where each woman has an average of just one child, is itself an extreme case.

In fact, precisely where the total fertility rate (the total number of children born per woman) in Singapore stands is not entirely clear. The Singaporean government declares it to be 1.1, while other sources suggest 0.83.[5] A fertility rate of 1.0 is therefore a reasonable approximation. It must again be emphasized that when demographers talk of fertility, they are referring to *actual births*, not to the *ability* of women to have children. Women in Singapore are simply not having many kids, and whether this is due to problems with biological fertility or to a host of other factors, demographers refer to the outcome as 'low fertility'. A rate of 1.0 means that each generational cohort in Singapore is half the size of the one that came before it.

We are coming to see East Asian societies as having inherently low fertility, but this is a relatively recent phenomenon – there was a time when Asia was perceived in the West to be a continent of prolific childbearing and vast families. Early in the twentieth century, Europeans spoke fearfully about the 'Yellow Peril', which had much to do with the size of Asian populations. China was helpless in the face of European incursions and Asia as a whole had fallen under European domination, but Europeans felt threatened by the one advantage Asia still possessed: its immense numbers. This was even more the case after Japan's victory over Russia in the 1904–5 Russo-Japanese war. The impression of Asia as a rapidly growing continent of teeming millions continued well into the middle of the century, even if the racial language became more tempered.

Back in the early 1960s, Singaporean women still had more than five children, a similar number to their mothers and grandmothers. But from then, the drop-off was precipitous – by the end of the following decade, the average was fewer than two children per woman. What changed in Singapore was a rapid and extreme case of what has elsewhere led to falling and then low fertility rates: increased education, especially for women.

Educated women do not, on the whole, wish to have very many children, because they want to pursue their own goals and careers. They have the means and the knowledge to control their fertility with contraception. And when they *do* have children, they want to ensure that they get a good education and to help them on their way in life, so they concentrate their limited resources on fewer children. Urban, increasingly educated and progressively wealthier, Singaporeans have, over the past fifty years, come to tick all the boxes that result in small family size.

Because lower fertility tends to be adopted first by the more educated, it often leads to a worry that the 'wrong sort' are breeding too much and the 'right sort' too little, which jeopardizes the 'quality' of the nation. This fear was seen in the UK and Germany at the start of the twentieth century, and in Singapore at its end.[6] But what has made Singapore unusual is that its leadership was prepared to take action. Although British and German commentators worried about the quality of the nation's stock at a similar point in their demographic transitions, their governments were not prepared to act in a manner that favoured births in one class over another. But late-twentieth-century Singapore had a government that was far keener to intervene.

In 1990, Singaporean women who did not finish secondary school had one child more than those who graduated from university. A decade later the gap was half a child, as the less-educated began to catch up with their better-educated sisters.[7] Small family size is a social habit that is first adopted by the better-off and then percolates down through society. The worry is then not that the next generation will be of too low a quality, but that it will be too small in number.

Population policy has long been important to Singapore's ruling People's Action Party; initially, smaller family sizes were encouraged, with discrimination against large families in terms of housing and schooling.[8] When falling fertility among the best educated first began to appear in the early 1980s, the government of Lee Kuan Yew, the founder of the modern state of Singapore, promoted fertility among those he considered the best and brightest. In 1983 he stated:

> We must further amend our policies, and try to reshape our demographic configuration so that our better-educated women will have more children to be adequately represented. In some way or other, we must ensure that the next generation will not be too depleted of the talented. Government policies have improved the part of nurture in performance. Government policies cannot improve the part nature makes to performance. This only our young men and women can decide upon. All the government can do is to help them and lighten their responsibilities in various ways.[9]

A Singaporean friend of mine, who returned home in the mid-1980s after attending university in the UK and took a civil service job, was assiduously encouraged to attend 'love cruises', in the hope that she would meet an equally intelligent Singaporean man and create lots of babies with genetic material the state considered suitable. At that time, less-educated women were being offered financial incentives if they underwent sterilization.

As small family size spread down the educational spectrum, the government emphasis moved away from eugenics and towards raising fertility across the board. In the 1960s, the government had told citizens to 'Stop at Two', but by 1987 they were advising people to 'Have Three or More (if you can afford it)', and launching incentives and tax breaks for childbearing. A further series of initiatives followed in 2001, including a 'baby bonus' scheme that paid an annual cash sum for second and third children. In both cases, the results were short-lived and meagre.

The Singaporean case study illustrates that while governments might be able to reduce birth rates effectively, increasing them is much harder. Whatever the government had done, given the speed of Singapore's economic and human development by the 1980s, Singaporean women would not have had five children each, as they had twenty years earlier. In trying to lower fertility between the mid-1960s and early 1980s, the government of Singapore was working with the flow of history, but trying to raise fertility was akin to pushing water uphill.

As of 2017, around 70 per cent of Singaporean women in their mid- to late twenties were single, up from just over 60 per cent just a decade earlier. The figure for men in this age group is over 80 per cent.[10] This is particularly important for fertility in a country where just 2 to 3 per cent of births occur out of wedlock, well below a tenth of the level in the UK, Germany and the USA.[11] If Singaporean women are neither marrying nor having children outside wedlock, they are not having many children.

Although fertility has declined faster and further in Singapore than in the USA, Canada and most European countries, it has nevertheless followed the same path. By the 1930s, women in many parts of Europe were having an average of just two children. After the Second World War, fertility rates rose with the famous Baby Boom, but by the 1960s they were falling again, with the advent of the Pill and changes in social attitudes. Just like the Singaporean editor we have already met, women in the West seem to have been put off having large families by the drudgery it involved, which conflicted with their rising education and expectations. Here, too, dirty nappies featured prominently among the factors that discouraged educated and ambitious women from having children: 'Years of scraping shit out of diapers with a kitchen knife, finding places where string beans are two cents less a pound.'[12]

### *The Infertile Crescent: A Journey Through the Lands of Baby Dearth*

If the low fertility rate of Singapore were an exception, there would be little to worry about. Singapore is, after all, a small country. An appealing destination for the populations of less developed and well-organized neighbours, it can avoid population decline by increasing the rate of immigration. It is even able to select immigrants who reflect its predominantly Chinese ethnic composition, either from neighbouring Malaysia, with its numerous Chinese minority, or from China itself. Indonesia is another potential source of ethnically preferred potential immigration. In this, Singapore is relatively unusual. Countries like the UK and USA can attract immigrants easily enough, but mass immigration inevitably means ethnic change.

In its low fertility rates, however, Singapore is in the widest international company.

It turns out that there is an *infertile crescent* that stretches from Spain to Singapore. You could walk from the Straits of Gibraltar to the Straits of Johor at the opposite end of the Eurasian landmass and barely set foot in a country with a fertility rate above the replacement level of around 2.2 children per woman.

The route would start in Spain, which has a total fertility rate of 1.3. Spanish women have been having fewer than two children since the early 1980s, and by the late 1990s the fertility rate was barely above 1.0. Even under the pro-Catholic and pro-natalist regime of Franco, fertility rates were not particularly high. There has been a small recent increase to a little above 1.3, but this can at least partly be explained by something known as the tempo effect. During the period when women are becoming mothers at a later age than previous cohorts, fertility rates are artificially depressed. When this slows or ends and the average age of the childbearing mother is no longer increasing – or at least not at the same pace – the fertility rate tends to bounce back slightly, to correct for the previous downward distortion.[13]

Crossing over the Pyrenees to France, fertility rates are a relatively healthy 1.8, but still below the replacement level. During the nineteenth century, France did not experience the same sort of population explosion as Britain, Germany and then Russia. But when other European countries' fertility rates plummeted, France's did not, so its relative position improved. The fertility rate in France today is similar not to the ultra-low rates of southern and eastern Europe, but of the more buoyant levels in the British Isles, Benelux and Scandinavia. Fertility rates are higher in these societies partly because women have been more successful at combining family and career, and because births outside marriage are more accepted. Immigration from countries with higher fertility rates has also maintained the fertility rate in countries like France, although immigrant birth rates tend to converge on those of the host country fairly rapidly.

From France, we move on to Germany and Austria, both of which have had fertility rates of no more than 1.5 children per women for several decades. Only mass immigration prevents these countries from having declining populations. Croatia, Serbia, Romania and Ukraine are next, and fertility rates in all these countries are below 1.75 children per woman, as indeed they are throughout southern and eastern Europe. This is true of countries as different as Hungary, where fertility has been below replacement level since the early 1960s, and Albania, where women were having more than three children as recently as the late 1980s. What many low-fertility European countries have in common is a fatal combination of laudable female educational opportunity and traditional values. If you encourage women to get an education but frown on those who combine career and family, they will often opt for an interesting job rather than the joys of motherhood.

Another feature of low-fertility societies is the low level of childbirth outside wedlock. It is notable that where traditional taboos fade and births to unmarried mothers rise, fertility rates start to recover. In Hungary, for example, while the fertility rate has increased modestly, births outside marriage account for almost half of the total gain. It seems that a breakdown in traditional moral norms, rather than government policy, has increased fertility rates; the Hungarian government's incentives have focused on encouraging women to have a third child, but the modest uptick in the fertility rate is explained by more people having one or two.[14]

We move on to Russia, a country with sufficiently high fertility in the late nineteenth and early twentieth centuries for the population to grow in the face of revolution, civil war, world war, purges and famine. Yet this was changing by the middle of the twentieth century, and small Russian family sizes were a key factor in the Soviet military and economic decline.[15] Since the start of the current century, Russia's fertility rate has recovered from around 1.2 children per woman to nearly 1.75, but decades of low fertility mean that the number of young women bearing children is small and leave Russia struggling to avoid population decline. Back in 1914, the situation was very different: the prospect of seemingly endless Russian growth was one of the factors that prompted Germany to risk war before things got out of hand.[16] Today, although Russia's fertility rate (children born per woman of childbearing age) has improved, its birth rate (the number of children born relative to the population as a whole) has not.

The fertility rate of neighbouring China, the Earth's population giant, is barely above 1.5. Some demographers put it lower, closer to 1.2. And there is little sign so far that the relaxation of its one-child policy to a two-child policy in 2015 – further relaxed with permission to have three children in 2021 – is having much of an impact. This should come as no surprise, given the country's rapid urbanization, rising living standards and improving female education. As China's industrial revolution took off, people of childbearing age went to work in towns and left their children behind, which inevitably disrupted family life.

Reversing a low fertility rate in a now-developed society is proving difficult. The objective for most Chinese couples is no longer an extra pair of hands on the farm but to optimize their children's life chances by investing in just one or two. 'I value my daughter's all-round education and development, and the importance of spending time together,' says a twenty-six-year-old accountant, resisting her parents-in-law's exhortations to have further children. 'When I think about having to work, and the economic pressure, I think having one child is enough.'[17] The fearsome Chinese 'Tiger Mother' cannot be fully effective if she has more than one or two cubs, a feeling shared by mothers the world over. Women in China may be on the same path as their Singaporean neighbours, with incomparably greater global consequences.

In China and elsewhere, low fertility is self-reinforcing. A single child will bear the burden of looking after elderly parents, leaving less time to bring up a family of their own. Additionally, those who are not used to big families tend to lower their expectations of family size and are often content to stop at just one child. And finally, the economy becomes geared to cater for small families, making having more children increasingly inconvenient.

On our trip across the Eurasian landmass, the highest fertility rate is found in Myanmar, where it is, at the time of writing, around replacement level. In the late 1970s, Burmese women were having more than five children, but fertility rates have since tumbled. The last countries before we arrive in Singapore are Thailand (with a fertility rate of just over 1.5) and Malaysia (just below 2). Thailand is a case study that shows how demographic change in relatively poor countries can outpace development – it reached the low fertility levels that were once associated with developed countries long before it was economically advanced.[18]

What is striking about this low-fertility 'world tour' is the varied nature of countries where a small family size is the norm. In European countries like Germany and Serbia, fertility rates have been low for decades, while in Asian countries like Thailand, the average woman was having more than five children as recently as the early 1970s. Some are rich, some are poor. Some are Christian, some Buddhist, some Muslim and others strongly secular. It is doubtful whether any of them will see significant increases in fertility in the foreseeable future.

The same low fertility rates persist in other, quite different, cultures. Lebanese women had an average of more than five children in the 1960s but today have fewer than 1.75. Almost nowhere has fertility fallen faster than in Iran. 'How can I even think of having children, when we can barely make ends meet?' laments one prospective Iranian mother.[19] And in Latin America, fertility rates are uniformly either low or falling. There are estimates that the fertility rate in South Korea, meanwhile, may now be as low as 0.8.[20]

The global baby dearth, then, is no respecter of regions or cultures. At one time, the Italian mamma was regarded as the archetypal matriarch, but that hasn't been the case for several generations. The only wide-scale exception today is sub-Saharan Africa, where declining mortality rates combined with the persistence of high fertility is fuelling history's greatest ever population boom.

### *Religion and Fertility: Toward Post-Modern Demography*

A country's level of development used to indicate its fertility rate and life expectancy. In the early 1970s, wealthy North Europeans lived into their seventies, while poorer South Asians failed to make it into their fifties. Today, although the development gap between the two regions remains, the difference in life expectancy has narrowed. The same is true when we compare the birth rates of countries on the other side of the Atlantic. Back in the 1970s, Brazilian women had more than twice as many children as American women. Today, although Brazilians are still barely a fifth as wealthy as Americans, they have slightly fewer children. Poor countries used to have big families and low life expectancy, but no longer.

On the surface, it might look as if we are reaching the end of demographic history, with one dramatic African chapter left to play out. In pre-modernity, all countries had high fertility rates and short life expectancy. Although there were variations – eighteenth-century England and Japan seem to have modestly curtailed their fertility, for example – the differences were not great. The only exceptions were caused by disasters like plagues and wars, when mortality was extraordinarily high, or periods of good harvests, when it fell a little. At different times and at different rates, the world has passed through a demographic transition. Once that is through, the key demographic difference concerns fertility rates, and the determinants are cultural rather than economic.

This can be seen if we compare fertility rates in different US states. South Dakotan women have around three-quarters of a child more than women in Vermont; the pro-natalist values associated with the American heartland are in stark contrast with the secular liberalism of New England. The more religious a US state, the higher its fertility rate, but conservatism correlates even more strongly with high fertility. At the state level, religiosity and a tendency to have voted for Donald Trump in 2016 are, respectively, twenty-five and forty times more closely correlated with fertility than income.[21] In Utah, the influence of religion on fertility is clear: Mormon families are famously numerous. As one Mormon mother of six says, 'People often asked me why I had so many children. I usually told them in a few words about the plan of salvation.'[22] The Amish of Pennsylvania, Ohio and Indiana have an average of five or six children, equivalent to the fertility rate in Niger and Chad. Their numbers grew from 6,000 to over 300,000 between 1901 and 2010,[23] an increase that was overwhelmingly achieved through fertility rather than conversion.

In the long term, this could affect the ethnic makeup of the USA in an unexpected way. The only two American states with above-replacement fertility, South Dakota and Utah, are both more than averagely white. Although fertility rates among white American women are still slightly lower than those of black or Latino women, the gap has narrowed. Latinos come from countries with traditionally high fertility but live in low-fertility urban areas. They have seen their fertility rates tumble towards local norms, which increasingly resemble the new norms in their home countries. The fertility rate in Mexico is now close to replacement level, with each woman having had an average of nearly seven children in the early 1970s. As recently as 2007, Hispanic fertility rates in the USA were about 60 per cent higher than rates among white women; today, three-quarters of that gap has evaporated.[24] Latino women were by 2016 having slightly fewer children than predominantly white rural women, and the gap here is only widening.[25]

Meanwhile, ultra-high fertility is found among predominantly white rural groups such as the aforementioned Amish and the Hutterites, who increased in number from 400 to 50,000 between 1880 and 2010, a rate of about 3.8 per cent per annum.[26] These groups are still relatively small and may not yet be much-noticed, but in the unlikely event that their expansion is a third of the way through its course at this rate, they would number half a billion by 2060.

Our assumption of low fertility and demographic shrinkage in white communities is based on a century of experience, but it need not always be the case. We might see urban areas in America and even Europe that are predominantly populated by people of mixed-race and non-European origin retreating in the face of a demographic resurgence from a rural white population. The signs may be slight, but they are there.

## *The Jewish Exception*

The ultra-Orthodox Haredi Jews are another demographically expansionist white American group. Unlike most other high-fertility American minorities, they are almost exclusively urban and they live in states where fertility rates are generally low. In the Brooklyn neighbourhoods of Williamsburg and Borough Park there are Haredi communities of tens of thousands, with family sizes that are similar to the highest-fertility countries in the world. They are growing rapidly, with few signs so far of a slowdown, which inevitably leads these communities to seek new space in new neighbourhoods.

Haredi communities in the UK are also growing by 5 per cent each year, at which rate they will double in around fifteen years.[27] Here, too, there is immense pressure on housing, so Haredi groups are beginning to establish satellite settlements beyond the inner suburbs, where housing is more affordable. This pattern is already established in the USA. A group of Haredim moved from Williamsburg in the 1970s to a new town called Kiryas Joel, which now has around 30,000 inhabitants and a median age of around thirteen (compared to thirty-seven in America as a whole).[28]

Haredim are also contributing to demographic growth in Israel. As is usually the case with groups that have big families over a long period, they are a young community – almost 60 per cent of them are aged under twenty, compared to 30 per cent in the rest of the Jewish population.[29] But it is not just the Haredim who have many children in Israel; fertility is higher across the religious spectrum than we might expect in such an advanced country. Israeli women have nearly three times as many children as Singaporean women, although they are no less educated. In Israel, as in America, politically conservative groups have more children, even if they are not religious.[30]

The emergence of values-based fertility will turn many of our demographic assumptions on their heads. We might expect Jews to have smaller families than Arabs and other peoples who are predominantly Muslim, but that simply isn't the case. In the early 1980s, Israeli women were having less than half as many children as Iranian women, but today they are having significantly more.

But the picture regarding the world's Jews is not uniform. In Israel they have a high fertility rate, with even the most secular communities at least above replacement level, but secular Jews in the United Sates have one of the lowest fertility rates of any group.[31]

Some demographers have referred to a second demographic transition to universally low fertility rates, as individualism replaces aspirations for a family,[32] but the existence of a universal trend has been exaggerated. Instead, fertility is increasingly linked to attitudes, ideology and religion. As in Christian and Jewish communities, religious affiliation and practice are associated with larger families in Muslim countries.[33] Some sub-groups within societies are choosing to have big families, while others prefer small ones; this will shift the demographic balance within and between countries. The Mormon population has grown fifteen-fold since 1947, in part through high fertility rates. If the offspring of those groups that opt for high fertility continue to adopt traditional attitudes to fertility, we might wonder whether secular societies might fade away, leaving the religious to inherit the earth.[34]

However, none of this is guaranteed – if religious groups are to keep growing, the retention of believers is just as important as their fertility rate. Here the data is limited. There is a definite drift away from Haredi lifestyles, both in Israel and outside, but it is almost certainly small compared with their natural growth.[35] At a societal level, there is a continuing move away from religion, both in the USA and most of Europe. Just as pre-modern cities required an endless migration of people from the countryside to replenish their populations, modern secularism attracts people from more traditional communities with high birth rates, who then fail to replace themselves. The future, it seems, belongs to those who can retain a culture of high fertility while also limiting defection.

### *Local Heroines and Eco Warriors*

One Friday morning I persuaded two friends of mine to join me to discuss demography. I was keen to talk to them because they contradict the rule that women have fewer children as they become more educated. Sarah, a graduate of Cambridge University, has six children. Vicki, an Oxford graduate, has seven. I wanted to understand their motivations.

Both Vicki and Sarah are Jewish and Orthodox; they are both modern in their outlook. More than any strict sense of religious obligation, I felt that their large families were explained by a love of children that fitted into a pro-natal culture. Vicki edits a community newspaper from home, while Sarah, who was a lawyer before her children were born, does not work outside the home. But both of them, intelligent and highly educated as they are, feel that motherhood is the most fulfilling role they could play. 'Bringing seven children into the world and sending them off as rounded, mature and responsible people is the most creative and fulfilling thing you can do,' Vicki said.

Neither wished to condemn those who chose to have smaller families, and both were sensitive to people who find themselves unable to have children, but when talking about the falling fertility rate in society, they could not help mentioning the word 'selfishness'. It was not that having lots of children in the modern urban world was necessarily very difficult, they said. It was occasionally hard to find a big enough hire car on holiday, or to book family tickets for events, but these were minor inconveniences. For Vicki and Sarah, people who chose to have fewer children were putting personal development, holidays or the ability of each child to have their own bedroom above the creation of new life. Neither Vicki nor Sarah wanted to condemn anyone for the choices they made, but they explained that we live in a world where achieving personal goals and a certain standard of living have become norms that are difficult to reconcile with having a sizeable family. Ultimately, such expectations may be impossible to reconcile with having a family at all.

Sarah did raise the possibility that she was being selfish in having a large family. There are growing concerns in the West that children are an indulgence, especially in the developed world, where consumption and emission levels are high. It is question that the US congresswoman Alexandria Ocasio-Cortez has raised: 'It is basically a scientific consensus that the lives of our children are going to be very difficult, and it does lead young people to have a legitimate question: is it OK to still have children?'[36] This is a subject to which we will return later.

## *The Future of Sex*

I recently sat on a panel for a seminar on demography, alongside an eminent retired professor of the subject. Referring to the question of low fertility, he referred to 'a reduction in coital frequency'. The audience giggled at his charmingly old-world and academic way of talking about something people find endlessly fascinating. We might usefully ask to what extent how much sex people are having affects population – while the overwhelming majority of births continue to result from sexual intercourse, it is a subject that no study of demography can ignore.

Increasingly, there is evidence that young people in countries as different as Italy and Japan are less keen than their predecessors on long-term relationships, marriage and childbearing. It even seems that they are less interested in sex. A quarter of Japanese adults under forty had never had full heterosexual sex and the proportion is rising.[37] In Italy, reducing levels of sexual activity have been attributed to a decline in men's sex drive.[38]

Japan may be ahead of the curve in its progression towards a future with less sex, but it is not alone. Lack of sexual activity among American millennials is twice the level of the prior generation, while the sale of condoms is on a steady downward path.[39] The rules of engagement in the #MeToo era can be confusing and off-putting, while a blurring of traditional gender roles may also be contributing. While increased male participation in household chores is laudable, it does seem to be associated with a reduced sex life.[40] The development of technology may be another factor. 'The reason the teenage pregnancy rate is now so low compared to when I started in my career,' a GP friend of mine told me, 'is because young people are all staying in their bedrooms with their technology rather than going out and having relationships.'

A decline in sexual activity need not, in itself, reduce the fertility rate, just as arriving late to sexual activity and marriage doesn't necessarily mean an individual will have fewer children. But the longer a woman leaves childbearing, the less fertile she is likely to be; at a societal level, later childbearing means smaller families. If you remove women who are in their peak fertile years from the pool of potential mothers, it is highly likely that at the aggregate level there will be fewer pregnancies and births. Communities in which women delay their fertility tend also to be those in which they are pursuing other projects than maternity and in which fertility is low for social reasons.

Often people mention television as a cause of low fertility, and there certainly is a modest fall in sexual activity that accompanies the arrival of television.[41] However, TV does not only lower fertility rates

## Total Fertility Rate of Selected Countries, 1950–2100

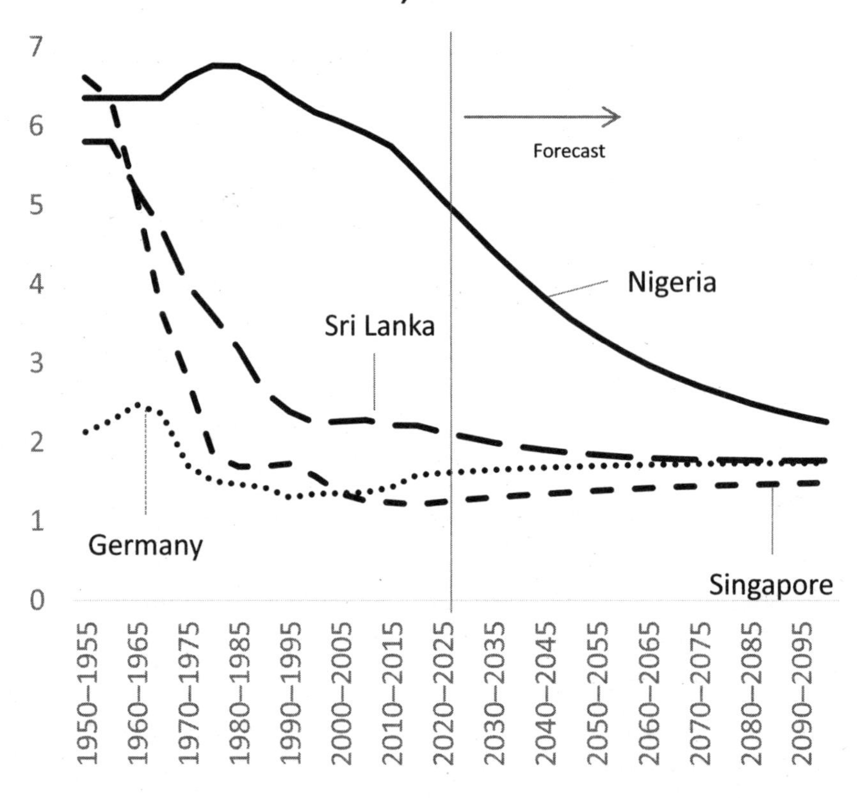


Source: United Nations Population Division (Medium Variant Forecast)

In the mid-twentieth century, there was a huge disparity between countries' fertility rates, with women in Africa and Asia tending to have six or seven children, while women in Europe and North America had two or three. Since then, the fertility rates in countries that have grown rich, like Singapore, and those that were already rich, like Germany, have plummeted to well below 2. Even relatively poor countries like Sri Lanka have seen fertility rates quickly fall towards 2.

Nearly all the countries with high fertility rates today are in sub-Saharan Africa, and demography's biggest unknown is how quickly they will fall. The UN expects a steady decline in Nigeria, Africa's population giant, with fertility rates only approaching replacement level towards the end of the twenty-first century.

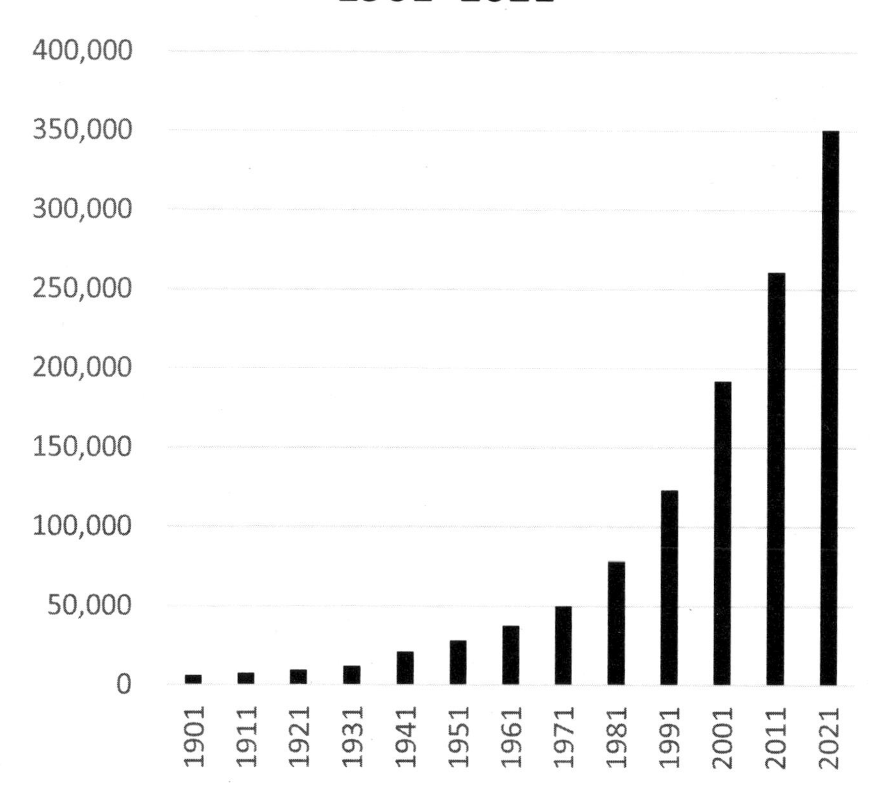


Sources: American Experience, Amish Studies
Note: Some data points smoothed

When a small population grows at about 3.5 per cent every year, its growth may initially go unnoticed. But eventually, if the trend continues, it will be big enough to have an impact on society. This is what has happened with the Amish community in North America. There were 6,000 of them at the start of the twentieth century, but today there are a third of a million.

Religious groups that can maintain high fertility and retain most of their members stand to become increasingly prominent in their home countries.

by distracting people from making babies. A Brazilian study found that it was the popularity of aspirational soap operas that lowered fertility rates; the luxury apartments, smart cars and stylish clothes glamorized the lifestyle of people without large families and caused people to have fewer children.[42] Big families come to be associated with the backward, rural ways of life that people in developing countries are trying to escape. Another twist, which we have already noted, is that the societies with the lowest fertility rates are those that combine an aversion to childbirth outside marriage with progressive attitudes to the education of women. In places like Britain and Scandinavia, where many births occur outside wedlock, the situation tends to be more rosy – at least if you like babies.

It may be that one's attitude to fertility has a genetic component.[43] In a world where people lacked choice, a genetic predisposition to want to reproduce might make little difference. But once people can control their own fertility, it could be that child-favouring genes proliferate, which would eventually mean a recovery in fertility rates.

### *Empty Planet?*

Thomas Malthus was wrong when he asserted that population pressure would forever be limited by resources. And demographers were wrong when they suggested that fertility would settle everywhere at just above two children per woman, allowing for a global population that was more or less stable. A baby boom in the West was followed by a baby bust, after which families in other areas shrank to the extent that population decline became a possibility. Sub-replacement fertility spread from western Europe and North America to southern Europe, the former Communist bloc and East Asia. Once the preserve of the prosperous, low fertility is now so widespread that the link to economics has been greatly weakened.

The countries that embraced small families earliest now have fertility rates that are moderately sub-replacement, while some later adopters have seen fertility rates plummet more dramatically.

We should be careful with forecasts, but some predictions can be made with confidence. The big uncertainty, as we have already noted, concerns sub-Saharan Africa. Provided that the continent continues to develop, its demography is likely to follow the rest of the world and decline in fertility. Even if its development is arrested, fertility decline is possible, as we have seen in the Middle East, though at a rate that is difficult to predict. Outside sub-Saharan Africa, we can expect the fertility rates in countries where it is high to continue to fall sharply. Once a country has a fertility rate of below four, the drop tends to continue, but at what level it will settle is uncertain. The fertility rate in Sri Lanka has been between 2 and 2.5 for almost thirty years. Compare this with Colombia, which spent barely a decade in this 'Goldilocks zone' and is now below replacement.

Whatever happens to global fertility, demographic momentum will ensure that the world's population will keep growing for the time being, albeit at a slowing rate. Nevertheless, we have already reached what the Swedish statistician Hans Rosling called 'peak child', the moment at which the number of children in the world stopped increasing.[44] By the end of the current century there are likely to be 50 per cent more people on the planet, but there will be more than fifty million *fewer* who will be aged under five.[45]

Fertility is the most important motor for changing demography. In principle, there is no limit to how low it could go – we might one day think of Singapore's current rate as being quite high. We tend to regard certain cultures or societies as marked by perennially high fertility, but we are usually wrong. Although India's fertility rate has only recently reached or just dropped below replacement level, several Indian states have a lower rate of around 1.7, and it is likely that the country as a whole will follow. Falling fertility means poor countries are progressing through the process of demographic modernization at a pace that is well ahead of their economic development.

India's fertility decline was later and gentler than China's, and the former is about to overtake the latter as the most populous country on Earth as a result. By the end of the current century, China is predicted to have lost around a quarter of its current population, causing worry about a shrinking workforce, while India is in full economic stride. The healthier state of India's demography will provide an opportunity to make up the ground it has lost to its rival in the past decades. In 1980, China's economy was a little more than one and a half times the size of India's; by 2016, it was between four and five times the size.[46] We can expect that trend to be reversed in the coming years, and much of the reason will be India's growing workforce.

Japan has long suffered from low fertility rates and inevitable economic stagnation, a classic case of the low-fertility trap. Where women are given educational opportunities, their fertility rate will usually fall to around replacement level, and where they are not encouraged to combine work with motherhood, their fertility rate will plunge further. Japan is full of women who are fulfilled neither as mothers nor as workers. It is no wonder that the Japanese are among the least happy people in the developed world,[47] despite their comfort, affluence and the country's lack of crime.

The possibility remains that the world could 'go Japanese', with Africa rapidly following other continents and countries like Sri Lanka also shifting down a gear. The argument is that people everywhere will eventually become educated and prosperous, leading to universal sub-replacement fertility, with men and women unprepared to spend the time or money necessary to raise a larger family.

Yet as we have seen, a post-modern fertility pattern is already emerging. Conservative values and religious faith invariably accompany higher fertility, whether that means a rate that is just above replacement level or the sky-high fertility of the Pennsylvanian Amish. We might see a world in which only groups with strong pro-natalist ideologies survive, while others fail to breed and simply disappear. In which case, we are heading not for an 'empty planet',[48] but rather a world filled with diverse groups that have sociological similarities but ideological differences. Were high-fertility groups, which often have absolutist ideas and a dislike of modern technology, to become dominant, the political and technical challenges of running a modern society would become even greater. While Japan may be a pioneer of demographic modernity, Israel could be the pioneer of post-modernity.

In the meantime, even before population decline sets in, low fertility rates lead to an ageing society, and it is to this phenomenon that we now turn.

# 5

# Ageing

## *43: The Median Age in Catalonia*[1]

Roussillon is an area of Catalonia that is in France rather than Spain. A land of snow-capped peaks and glistening Mediterranean coves, with vineyards that roll down to the sea, it is one of the few places in Europe where you can ski in the morning and sunbathe in the afternoon. Bounded by the sea and the Pyrenees, it has been a key crossing point between France and Spain for centuries. It is a place I am fortunate to visit frequently.

Over the centuries, Roussillon's Pyrenean mountain passes have served as a route for people and goods being smuggled in both directions. In 1939, hundreds of thousands of Spaniards headed north to avoid Franco's clutches; a few months later, refugees fled desperately in the opposite direction to avoid the Nazi invaders and their Vichy allies.

A couple of years ago I decided to visit Portbou, the first coastal town over the border in Spain, where the German-Jewish philosopher Walter Benjamin committed suicide in September 1940. Today it is a pleasant seaside town, although less significant than resorts like Collioure to the north or Cadaqués to the south. There is little to see besides Benjamin's grave and a memorial to him on the coast.

Although my visit to Portbou was more of a pilgrimage to the shrine of a dead philosopher than a demographic expedition, it helped to unlock something of a mystery. About a year earlier, the regional authorities had called a referendum on Catalonian independence and the people had voted in favour. However, Madrid refused to recognize the referendum result – it was clear that Spain was not going to allow Catalonia to become a separate state. There were scuffles on the streets of Barcelona and some injuries, but nobody was killed. There were no hot-headed attacks on rural police stations and no brutal retaliations from the army. Instead of the referendum leading to civil war, the story seemed to fade away and then disappear from the headlines altogether.

As I sat in one of Portbou's squares, I reflected on why the referendum had become a footnote in history rather than the trigger for violent conflict. I looked around at the grey-haired locals enjoying the October sunshine and sipping cups of black coffee. They were far too old to take up arms and march in the streets, enraged by political injustice. And although these seaside dwellers were older than the average Catalan by a couple of decades, even people in their forties do not tend to take up weapons in a political cause. On the whole, the middle-aged are too busy worrying about the health of their ageing parents, how their children are doing at school and how they're going to pay for their own retirement while keeping up with their mortgage payments.

The overall age structure of a society has an influence that we either take for granted or tend not to think about. I was struck by the same phenomenon when the people of Hong Kong took to the streets in protest at their government's concessions to Beijing's demands regarding extradition in 2019 and 2020. Although the protests engaged a wide range of the population, they were led by the young. Deaths were in the low single figures,[2] but it was a completely different story from the Tiananmen Square protests in Beijing in 1989, when as many as ten thousand people were killed by the Chinese army.[3] If we contrast the median age of China at the end of the 1980s (around twenty-five) with that of Hong Kong three decades later (nearly forty-five), we have the beginning of an explanation. We have to assume that the Chinese authorities would have done whatever was required to preserve order in both cases, but without a swelling tide of young people in the population, what was required was lower in scale and the result was less deadly. We have heard about countries getting old before they get rich. There is a danger they may get old before they get free, and lacking risk-taking and enraged youngsters, put up for ever with authoritarian regimes.

### *War and Peace, Young and Old*

In Catalonia today, people in their forties by far outnumber those in their twenties, while the median age is well over forty.[4] The first Catalonia I encountered was in George Orwell's *Homage to Catalonia*. Far from the sun-kissed valleys and snowy peaks that I'm familiar with in Roussillon, Barcelona in the 1930s was a chaotic wartime urban melee. Back then, the city was run by a revolutionary mix of Communists and anarchists and the median age in Spain was barely half what it is today.[5]

In undergoing this change, Spain has typified many countries that have passed through the demographic transition. Decades of sub-replacement fertility rates and lengthening life expectancy (now at over eighty-three, one of the world's highest) have resulted in a population that is exceptionally old, both by its own standards and by international comparison. As a country develops this is almost inevitable, except for places where a strong ideological or religious motive resists the tendency to smaller family size. So the age structure in Spain and similar European countries like Germany and Italy is a precursor of what is in store for most of humanity.

Ageing is followed by a vast increase in the very elderly and, without mass immigration, by a shrinking overall population. These are topics for future chapters, but here I want to examine what the upwards shift of the median age means for a society and the world as a whole.[6]

It stands to reason that a society where the average person is in their early twenties and the majority of people are youthful will be different from one where the median age is in the forties and young people are thin on the ground. Just as we expect the atmosphere in a nightclub to be different from that of a cafe, we can expect a society where the young dominate to be different from one that is middle-aged.

If societies that are heavily weighted towards the middle-aged and elderly differ from those full of young people, nowhere is the difference more significant than in matters of conflict. We have already seen the difference between the young Catalonia of the 1930s and the older Catalonia of the 2010s, where not a single life has been lost in a political struggle for independence. It is demography that explains why Catalonia was plunged into civil war in the 1930s and why it was not in the 2010s.

It might be pointed out that while Catalonia has remained at peace, this has not been true of the Basque country at the opposite end of the Pyrenees, but the conflict there reinforces the same point. In the 1960s, when Basque nationalists initiated violence, the median age in Spain was barely thirty, around fifteen years lower than it is today. But as Spain aged, the energy drained from the conflict, and a ceasefire was called in 2010. Ageing also sapped much of the energy from the conflict in Northern Ireland. This is not to discredit the diplomatic efforts of the politicians who crafted the peace agreement, but they did have the advantage of working with the grain of demography. Today the average age in Ireland is nearly forty, having risen from the mid-twenties since the mid-1980s.

If we take a closer look, we can find many examples of violent conflicts losing their energy as populations aged. When civil war broke out in Yugoslavia in the early 1990s, the median age in Bosnia was less than thirty; today it is over forty. In Serbia, too, the median age has risen by almost ten years in that period. However uncomfortable and ambiguous the constitutional settlements in Bosnia and Kosovo are, peace has prevailed for more than twenty years.

The Middle East also provides several good examples of this trend. When civil war began in Lebanon in the mid-1970s, the median age was in the late teens. A generation or so later, the country has managed to avoid another slide into open conflict. Although street protests have turned violent, the number of deaths at the time of writing can be counted on one hand. This is partly attributable to the fact that Lebanon's median age is close to thirty, and rising quickly. There has been instability, with a financial crisis and the devastating explosion in Beirut in August 2020, but civil peace was maintained at least until the autumn of 2021, thanks in part to the country's age structure. Syria, meanwhile, where the median citizen is still very young, has not been spared the calamities of war and massacre.

The claim that older societies are less prone to war is not just supported by anecdotal evidence, but by serious statistical and scholarly work.[7] In the 1960s, it was observed that the rise of the Nazis had been accompanied by a surge in the number of young men in the German population. It seems that Europe's instability in the first half of the twentieth century, followed by a long period of peace, can to some extent be explained by youthfulness being followed by ageing – the continent's median citizen is now more than a decade older than at the end of the Second World War. Studies of decades-long periods reveal that there is almost no civil war in countries where 55 per cent or more of the population is aged over thirty.[8]

In the modern era, young populations tend to be poorer, and

poorer populations tend to be more violent. We might think that young populations are violent because they are usually poor, but there is a close relationship between the relative size of young versus older cohorts and the tendency for civil war – by far the most common kind of conflict – to break out.[9] It is hard to imagine the Rwandan genocide of 1994 occurring in a country where the median age was in the forties rather than just eighteen years old.

It is not just war that is related to the youthfulness of a population, but crime. Before we examine the further ways in which demography impacts society, it is worth asking why this link exists and how it operates.

### *Explaining the Link*

While it cannot be said that youthfulness 'causes' war, or that maturity 'causes' peace, a society's age structure creates background conditions against which other things either do or don't spark conflict. The Catalonian independence referendum or the Hong Kong extradition treaty provide a spark that's much more likely to cause an inferno when a sizeable cohort of young people is there to catch fire. However, where the surrounding human material has been damped down by the predominance of older people, it may simply peter out. The sort of massacres suffered by Rwanda in the early 1990s just don't happen in places like Portbou.

What matters is not the number of young people in a country, but the relative sizes of younger and older cohorts.[10] In absolute terms, there are many more young people in Germany than in Guatemala, but that is not the point – Germany has a much larger population. What makes Guatemala more violent than Germany, at least in terms of how age explains the difference, is that Guatemala has around two twenty-somethings for every forty-something. In Germany, by contrast, there are almost 50 per cent more people in their forties and fifties than there are aged under twenty.[11] This is why the median age is such a useful measure; it locates the demographic centre of gravity in a population, and the older it is, the more stable a society tends to be. Older people, it seems, are a constraint on their juniors, while the absence of such a constraint allows youthful hot-headedness to set the tone.

This demographic centre of gravity also seems to explain something about culture. Nightclubs in the UK were shutting down long before the Covid-19 pandemic.[12] If younger people are having more early nights, drinking less alcohol and having less sex, it might be because their relative lack of numbers means that they no longer dominate the culture. Instead, they are influenced by their more numerous elders. The era of youth culture, it seems, was born with the baby boomers and has faded with their ageing.

However, this does not explain why youth is associated with violence and war. The link of youth to turbulence, and middle and old age to social calm, is something we tend to take for granted, but there are two quite convincing answers that have to do with intergenerational biological and social differences.

The human brain changes biologically between puberty and middle age, for good evolutionary reasons. It can hardly come as a surprise to parents of teenage children that adolescents tend to be moodier and more impulsive than adults. Scientists tell us that high levels of testosterone, oestrogen and progesterone course through the adolescent body, resulting in emotional and volatile responses.[13] Young people are influenced more by other young people than by the restraining guidance of their parents.[14]

These traits tend to make teenagers more violent and prone to take risks. Young male drivers are six times more likely to be involved in serious road accidents than their share of the UK driving population would suggest.[15] The markedly higher chance of a young man being responsible for a car accident[16] is not just a case of relative inexperience, but of a chemical and biological difference in their brains, which affects their split-second decision-making. The higher insurance premiums charged for young drivers, far from being capricious, are based on real calculations of risk.

There is also the basic biological point that people tend to weaken as they age. In a fight between one individual at the peak of physical prowess and another approaching middle age, the former is at an advantage. Once people have passed their mid-thirties, they find ways to manage conflict other than by the use of force, where they are at a distinct disadvantage.

During one's twenties, the regions of the brain's frontal lobe associated with self-control, judgement, planning and risk continue to develop. The result is that a thirty-year-old is less likely than a twenty-year-old to take the sort of impulsive and rash decisions that, when taken collectively, can turn a protest into a riot, and a riot into a civil war.

The other thing that changes as we move towards middle age is that we take on personal responsibilities. An eighteen-year-old rioting in the street might think they have little to lose; ten or fifteen years later, they will be weighing a more complicated set of personal concerns. Who will pay the mortgage if something happens to me? Who will put food on the table for my children? If I get arrested and end up with a criminal record, will I lose my job? At thirty, people are more likely to have put down roots and entered into long-term relationships. At forty, they are even more likely to have other commitments. Thus we can see that the link between youth and violence has a social as well as a biological explanation.

As people pass through early adulthood they are more likely to enjoy stable and sustainable sexual relationships, many of which involve marriage. In societies where pre-marital sex is frowned upon, many young people suffer from sexual frustration. In the Middle East, thanks to the expense of housing in increasingly urban societies, high unemployment and the cost of a dowry, young people form partnerships at a later age than in previous generations. The average Middle Eastern male does not marry until his early thirties, which is later than in almost any other world region. Coupled with taboos around extra-marital sex, the result is a silent, seething sea of sexual frustration that may well be partly responsible for many of the region's ills. As one commentator has written, 'Angry and with little else to do, many unemployed, highly educated and single youth turn to peaceful protest – or worse, militancy – and represent a legitimate threat to the security of Arab regimes.'[17]

The calculus of personal profit and loss changes as an individual ages, impacting their behaviour. As they approach middle age, people have more of a stake in the system. A breakdown in the social and financial order could lead to the loss of a small nest egg. War could lead to the destruction of physical capital in which the thirty-year-old might have a material stake, such as a home, a shop or a business. Young adults, by contrast, without any meaningful accumulation of capital or property, might regard a shake-up of the prevailing order as an adventure or even as an opportunity. So where the older generation predominates, it seems reasonable that we might expect more stability and less willingness to rock the boat.

## *Sons and Warriors*

There is another explanation for the reduced appetite for war among ageing populations: a higher median age is a result of smaller family sizes, and those people who have fewer children appear to be more reluctant to sacrifice them for a cause.

It might seem callous to state that people with many sons are prepared to give them up. It certainly doesn't correspond to my experience – as far as I can tell, those people with many children value each of them as much as parents with only one or two. But perhaps, where family size is *generally* small, the priorities of the majority with few children spread to the relatively small number of people with many. Where big families are normal, however, attitudes may be very different – there is no doubt that where the average woman has three or four sons, her protective instinct seems less prevalent and society seems more bellicose. Societies that shy away from war are not just those with the restraining influence of older populations, but those with fewer young people they are prepared to lose. The German theorist Gunnar Heinsohn suggests that where there is a sizeable cohort of youths, 'young men tend to eliminate each other or get killed in aggressive wars until a balance is reached between their ambitions and the number of acceptable positions available in their society'. In the late twentieth-century civil wars in Algeria and Lebanon, he argues, 'the warring stopped because no more warriors were being born'.[18] (We should remember, however, that it has often been old men who send the young ones into battle.)

It may sound strange to think that people respond to demographic stimuli as if they were rats in an experiment, but Heinsohn's theory seems to be playing out in places like Lebanon. In 2006, Israel and the Lebanese militia Hezbollah came to blows in a short, messy and inconclusive war. Today, Hezbollah is thinly stretched, having sustained significant losses during its battle with Syrian rebels over the past decade. It faces a recruitment shortage among Lebanese Shia Muslims, who are increasingly urbanized and producing fewer children. Indeed, while Lebanon's fertility rate was almost double Israel's in 1960, today it is barely half. To put it another way, when civil war broke out in Lebanon in the 1970s, the average twenty-year-old man would have been one of three brothers and one of six children. Today, by contrast, the average twenty-year-old Lebanese man is one of just two children.

This may well be a reason why Hezbollah has chosen to preserve the calm on its border with Israel for the past fifteen years, despite Israeli attacks in Syria. The ageing of the Lebanese population and the lack of young men ready to take up arms seems to be contributing to peace within Lebanon and its southern neighbour, while a change in the attitude of mothers who are more likely to have one son rather than three or four might also be a factor. An organization whose followers have a fertility rate of around 2 and which has suffered major losses in recent years can only fight so many wars. And just as an uneasy peace has held within Lebanon itself, so it has on the Israel–Lebanon border, at least up to the time of writing.[19]

At a global level, the fact that the United States is ageing more slowly than its rivals from the past (Germany and Japan) and the present (China and Russia) may extend its dominance. Whichever power comes out on top, the fact that *all* the world's major powers are ageing is one reason why there have been so few major conflicts in recent decades.[20] Indeed, some theorists have gone so far as to talk of 'Pax Americana Geriatrica'.[21]

### *Demographic Engineering: Population Strategies in Ethnic Conflict*

It is clear that war is more likely with a younger population and less likely with an older one, but the flow works the other way, too. In addition to demography shaping conflict, in conflict situations, the structures of populations can be changed by something called demographic engineering.

Demographic engineering is the intentional pursuit of demographic strength by ethnic groups in a state of conflict. They might do this as an end in itself, or as a way to achieve political or military power.[22] During a conflict, one group will try to strengthen its hand by manipulating events so that its numbers grow relative to those of its opponents. In the past, more people provided an advantage on the street or on the battlefield; in more recent times, a dominance of numbers means advantages at the ballot box, where power is now determined. This matters particularly where political parties run along ethnic lines, as is the case in much of the world.

Demographic engineering can be 'hard' or 'soft'.[23] The hard kind involves changing the demographic balance by directly impacting the population itself; the soft kind entails changing the demographic balance by other means, whether by shifting boundaries on a map or boundaries of identity. If this seems rather abstract, some real-life examples should help.

Take Romania under the Communist rule of Nicolae Ceauşescu between the 1960s and 1989. Keen to grow Romania's population, he was unhappy about the falling fertility rate, which was barely 2 by the early 1960s. As in Mussolini's Italy a few decades earlier, national targets were set for population size.[24] But at the same time, Ceauşescu wanted his country to be more ethnically Romanian. The great movements of people at the end of the Second World War had led to increased ethnic homogeneity in Eastern and Central European countries – millions of Germans were expelled from Poland and Czechoslovakia, for example.[25] However, Transylvania retained a significant Hungarian population, Romania had a relatively large number of Jews and Roma, and many towns and villages in the country were predominantly German.

So although contraception and abortion were banned in 1966, resulting in a spike in the birth rate, the Romanian authorities are said to have turned a blind eye to abortions in areas that were predominantly Roma or Hungarian.[26] More explicitly, ethnic Germans and Jews were allowed – for a price – to emigrate to Germany and Israel, providing the double benefit from the regime's perspective of foreign currency and consolidated ethnic demography. Thus, by encouraging selectively different birth rates and emigration, Romania's demography was engineered in line with the government's ethnocentric goals – reprehensible, but less so than hard demographic engineering's most brutal tactic: genocide.

The Romanian case is a clear example of hard demographic engineering, but in Northern Ireland, both hard and soft techniques have been used. On the hard side, the Unionist authorities were eager to ensure the maintenance of an approximately two-to-one ratio of Protestants to Catholics in Northern Ireland. To that end, they pursued housing and employment policies that encouraged the disproportionate emigration of Catholics, who also had an elevated birth rate. While this might be regarded as the result of religious practices, comparable birth rates among Catholics south of the border were lower, even though contraception was harder to come by.

The fact that the birth rate was higher among northern Catholics than those in the South suggests that the conflict was a contributing factor. Until the 1960s, higher Catholic emigration coupled with more Catholic births more or less cancelled each other out.[27] But once they stopped emigrating in such large numbers, their share of the population began to increase.

A case of soft demographic engineering, whereby the demographic balance is changed without affecting births, deaths or migrations, can be found in the creation of the Northern Irish state. When Ireland sought independent status, the UK government and the Ulster Unionist Party determined that the portion of the north of the country that could be separately retained within the UK should not extend to the whole of traditional Ulster. The counties of Donegal, Monaghan and Cavan were 'sacrificed' by the Unionists, in order to guarantee a sustainable Protestant majority in the North and to secure control over its Stormont parliament.[28]

The above case studies illustrate how population strategies have the potential to shape conflicts. A demographic perspective throws a new light on many conflicts, from the Middle East to South Asia, as well as in surprising places like the United States.

### *The Revolutionary Impulse*

Just as a young population is more likely to go to war, it is also more likely to engage in revolution. At the start of the demographic transition, populations tend to get younger because babies and young children who would previously have died early survive. We can see this happening in countries like Malawi, where infant mortality is between a third and a quarter of what it was in the late 1980s. The median age fell by a year or so, after which it started to rise as life expectancy increased. And as we saw in Chapter 1, plummeting infant mortality in the French territory of Mayotte led to a fifteen-year fall in the median age, and a similar effect was seen in the more distant past. When the demographic transition began in Europe in the late nineteenth and early twentieth centuries, the streets filled with young people who would previously have died in infancy.

Russia was a country of the young in 1917, and its revolutionaries were representative of their society: Lenin, their leader, was under fifty, and Stalin and Trotsky were both under forty. Further down the revolutionary hierarchy, many people in their twenties held positions of authority – they were a long way from the greying revolutionaries that would sit atop a greying society some seventy years later. Similarly in Iran, when the masses took to the streets in 1979, the median age was under twenty. If their revolutionary fervour seems to have waned, it might be because the median Iranian is now over thirty and ageing fast.

As societies age, it is not just political revolutions that become rarer; the same is true of cultural revolutions, and the same biological and social factors are responsible. So we should not be surprised that in the West, post-war unrest peaked during the late

1960s, when the first baby boomers were passing into young adulthood. In the era of Civil Rights, Vietnam protests and campus turbulence, the median American was below thirty; today, they are fast approaching forty. Campuses may still be hotbeds of dissent, but they are rarely centres of violent protest. Indeed, the retirement of the baby boomers has heralded other cultural shifts. Britain's *New Musical Express* closed its print version after sixty-six years in 2018, while cruises and glossy newspaper supplements on managing pension income have proliferated, the former at least until the coronavirus outbreak.[29] The cultural as well as the political centre of gravity seems to have shifted.

It was the same in China, where the flames of the cultural revolution more or less coincided with the youth revolts in Paris and Berkeley, at a time when there was a great bulge in the young population. Mao was deliberately appealing to a youthful audience, preserving his power by undermining the older party stalwarts. By the 1960s, thanks to a dramatic drop in infant mortality, the Chinese median age had plummeted from the early twenties to the late teens. Teachers and professors were beaten up and killed, and bureaucrats and officials were violently set upon in the name of revolutionary change, a movement that sought to destroy the 'bad old ways' represented by the middle-aged and elderly. Today, as in the USA, the median citizen of China is in their late thirties and ageing fast. Whatever else we can expect from China, we can be confident that there will be no more cultural revolutions.

## *Crime and Punishment*

The link between age and criminality is so obvious that we rarely think about it. When elderly people commit crimes – apart from those of the distinctly 'white-collar' variety – it still surprises us. In 2015, a group of men with an average age of sixty-three carried out

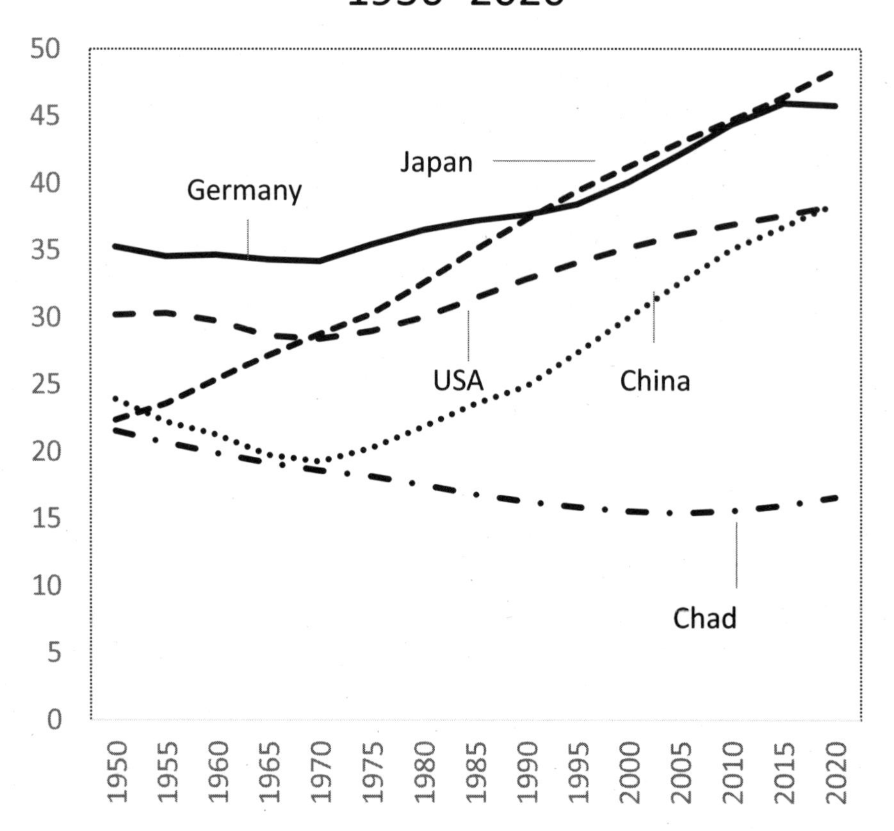


Source: United Nations Population Division (Medium Variant Forecast)

As fertility rates tumble and life expectancy lengthens, societies become older, and this is best seen by tracking a population's median age. In many European countries like Germany and some Asian countries like Japan, the median age is already over forty. And China, where the median age was barely twenty in the early 1970s, is fast catching up.

In less developed countries like Chad, the rapid decline in infant mortality while high fertility continues can mean that the median age falls as the country fills up with youngsters. But even here, as life expectancy gradually increases and fertility rates start to fall, the tide has started to turn.

a carefully planned heist in London's Hatton Garden jewellery district.[30] Even for a crime that required experience and meticulous planning, the age of the criminals was highly unusual and turned them into minor celebrities, with two films made about their exploits.[31] And when we hear of a city centre stabbing, we invariably assume that the assailant is male and in his teens or early twenties – and in making that assumption, we are invariably correct. Half of knife crimes in London are committed by teenagers or children.[32]

Just as with war or revolution, not everything about crime can be put down to age. If being young caused crime, the majority of young people in most societies would not be law-abiding. However, the tendency of the young to get involved in crime is strikingly disproportionate, for the reasons we have already discussed: a combination of biological impetuosity and having both relatively little to lose and something to gain.

Not all young societies are violence-ridden, but almost all violent societies are young. If we look at Bangladesh and El Salvador, we can see that they are both relatively young countries with very different murder rates – the rate in El Salvador is around thirty times that of Bangladesh.[33] But although not *all* young societies are violent, there are no countries with a high murder rate and a high median age. While most countries with older populations are rich, Bangladesh and plenty of others, from Malawi to Vietnam, prove that relative poverty need not result in violence. In fact, youth seems to be a far better predictor of violence than poverty.

Much of Latin America continues to be violent and is still young, but the continent is starting to age. In Mexico, the median age has risen from seventeen to twenty-eight in the past four decades, yet violence remains high. The authorities tackling the problem have got demography on their side, although their efforts are hampered by incompetence and corruption. Closer to home, it's notable that among the thirty-three London boroughs, the one with the highest murder rate has the second-youngest median age, while the two boroughs with the lowest murder rate are among the oldest.[34] What holds at the level of the individual and the nation is also true at the intermediate level of municipality or region.

Although knife crime in our major cities receives a lot of attention, much of the developed world has seen a drop in violent crime since the 1990s. A fairly well-developed literature attributes the drop-off crime in New York from the late 1980s to the liberalization of abortion laws and the consequent increasing prevalence of abortion, which removed many individuals who would have become criminals from the population.[35] Whether or not this much-debated Donohue–Levitt hypothesis is correct, some have attributed the decrease in crime to the reduction in young people as a share of society, while others claim that it has more to do with them being less inclined to commit violence.[36] Even if this latter theory is true, the restraining influence of the rising proportion of older people could be one explanation for the change. Furthermore, older societies are not just more peaceful and less crime-ridden; they are also more conducive to democracy.[37]

Sipping my coffee alongside the retirees of Portbou on a mild day in October and comparing the relaxed atmosphere with the tension I felt in Israel on the eve of the First Intifada, it didn't seem like somewhere that was about to descend into violence. As we move to an older world, we should move to a more peaceful one, with the general decline in violence encouraged by a tail-wind of demographic forces. However, a large part of the rise in median age has to do with the extension of life expectancy – and the continuing rise of the elderly as a share of the population is going to have a much bigger impact on the world than merely calming it down.

# 6

# Old Age

## *79,000: People in Japan Aged Over 100*

Seventy-nine thousand centenarians does not sound very significant in a country with a population of over 120 million people. Although this seemingly modest number is barely more than one twentieth of 1 per cent of the Japanese population, or around one in 2,000 people, it is almost certainly the highest share in any society in history.[1] And its scale tells us something important about Japan, which in turn tells us something important about the future of the world: that not only will median age increase, but the elderly will proliferate.

Nearly 90 per cent of Japanese centenarians are women. When Chiyo Miyako, the world's oldest person, died in July 2018, she was 117. She had inherited the title from Nabi Tajima when the latter died three months earlier. And after Miyako's death, the title passed to Kane Tanaka, a spring chicken at 115,[2] who continues to hold the title at the time of writing. By the time you read this, the baton may well have been passed on – in all likelihood to another Japanese woman.

Although extreme longevity is not entirely the preserve of the female sex, being Japanese helps even if you are a man. Masazo Nonaka was born in July 1905 and was recognized as the world's oldest man at the end of 2018, shortly before his death in early 2019. Already in early middle age when his nation plunged into war with the United States following the attack on Pearl Harbor, he was born during the Russo-Japanese War and died months before the accession of Emperor Naruhito. He put his longevity down to his fondness for bathing in hot springs and eating sweets.[3]

It should come as no surprise that the world's oldest women and men tend to be Japanese, but more and more countries are seeing growing numbers of people live to extraordinary old age, including an increasing number of men. As I was writing this chapter, my wife drew my attention to the death notice in our local parish weekly of a retired doctor who had died aged 105, mourned not only by his great-great-grandchildren but by two sisters. While the ages of Phyllis and Mirry were not disclosed, they can't have been much younger than one hundred themselves.[4] The oldest man and the oldest woman in the UK at the time of writing were, by sheer chance, both born on 29 March 1908.[5]

The world record for longevity is held not by a citizen of Japan but by a Frenchwoman, Jeanne Calment, who died in 1997 at the age of 122. Her claim has since been challenged by the Russian mathematician Nikolay Zak, who argues that Jeanne's daughter Yvonne swapped identities with her mother and died in 1997, sixty-three years after her mother had actually died.[6] If this is true, the record would pass back to Japan, but it is, in any case, significant that Calment came from the city of Arles in the southern French region of Provence. The Mediterranean basin is noted for the longevity of its residents, which is attributed to a diet rich in olive oil and light on animal fats – a more convincing explanation than Masazo Nonaka's fondness for confectionery. Indeed, two of the five so-called Blue Zones, areas of extraordinary concentrations of longevity, are in the Mediterranean: the Italian island of Sardinia and the Greek island of Ikaria. The other three are the Nicoya Peninsula in Costa Rica, Loma Linda in California and the Japanese island of Okinawa.[7]

*Life Expectancy and the Rise of the Centenarians*

Just as median age can tell you a lot about a society, life expectancy can tell you a great deal about its old people, especially once infant mortality and death in early or mid-adulthood grow statistically insignificant.

The way life expectancy is calculated was originally designed by the nascent life-insurance industry in the seventeenth century.[8] Life expectancy is essentially the reverse of the probability of mortality, being based on the statistical chance of surviving, or of *not* dying. As was explained in the introduction, it can be expressed at any age, but when we talk of life expectancy, we usually mean 'at birth'. In societies with high rates of infant mortality, a child who has made it to their first birthday will have a longer life expectancy than a newborn, because they have survived the risky first year of life. In most countries the number of years you can expect to live for simply reduces as you get older. The data is often quoted separately for men and women because there is a significant gap between the two, with women normally expected to live longer.

Globally, life expectancy has increased on average from the mid-forties to the early seventies since 1950, a stupendous and transformative achievement, but progress is patchy. Those countries that had the shortest life expectancy in the middle of the twentieth century have generally made the fastest progress, while life expectancy in the best-performing countries has increased from the mid-forties to the mid-eighties.[9] The Maldives, Oman and South Korea have extended life expectancy by *over forty years* since 1950, which in the first two meant that the length of the average life more than doubled. As with infant mortality and per capita income, there has been a great 'levelling up' in recent years, with the progress of the best countries slowing down while many previous laggards have made great headway.

This is part of the trend that we have already observed of convergence on the demographic norms of Denmark, with moderately low fertility rate, super-low infant mortality and generally long life expectancy. Let's compare Canada and Colombia, for example. Canada, already doing well in 1950, is doing even better, and life expectancy has extended from the late sixties to the early eighties. But Colombia, a much poorer country that started from a worse position, has made almost twice as much progress in the same period, with life expectancy increasing from the early fifties to the mid-seventies. The gap between the two countries has narrowed from eighteen years to just five, and the reason is that even where countries are still poor, priority is given to those resources and facilities that extend life. As we have seen elsewhere, the developing world has rushed through modernity, while the developed world has paused at the edge of what can be achieved for the time being, on the brink of a post-modern age.

Japan was already doing well in the post-war era, having recovered quickly from the disasters of Hiroshima and Nagasaki and its defeat in the Second World War. In the middle of the twentieth century, its life expectancy was already in the early sixties, only a little lower than in Canada and most of Western Europe. Since then it has successfully propelled itself into the front rank and now, along with Hong Kong, it has surpassed its nearest rivals. Early death in Japan became statistically insignificant long ago; the lengthening of life expectancy in recent decades has been all about the elderly.

When the elderly become extremely numerous, it profoundly influences almost every aspect of a society. There is a Japanese term, 'rougai', which is used to describe the elderly when they are annoying the younger generations, whether by obstructing closing doors on the Tokyo metro or offering unsolicited advice to the diminishing number of young mothers.[10] The term, as one foreigner living in Japan has observed, reflects the frustration of the working population at being outnumbered in a society they perceive as being cluttered by the retired.[11] It suggests a decline in the deference towards the elderly that was once a hallmark of Japanese culture.

Along with the increase in overall life expectancy, the growth in the number of super-old people in Japan is startling. As recently as 1990, the UN estimated that Japan was home to just two thousand centenarians, a far cry from today's seventy-nine thousand. Fifty years from now, the number will have increased *ten-fold*, while the population of Japan as a whole will have shrunk from over 125 million to less than 100 million, more or less where it was in the mid-1960s. While centenarians will not exactly be taking over, their numbers will be growing at a fantastic pace.

Like Spain, Japan's demography is post-modern, with fertility far below replacement level, but Japan has progressed even further because as recently as the 1970s the average Spanish woman was having almost one child more than the average Japanese woman. An advanced example of the low fertility that characterizes many highly developed economies, the land of the rising sun represents the future for much of the world.

The rise of extreme old age is not an isolated development, and should be looked at alongside the numbers highlighted by the preceding chapters. First a society prevents the mass death of its young, as in Peru, which causes its population to balloon, as is happening across Africa. Then many of the additional people, unable to make a living on the land, migrate to mushrooming cities, as we have seen in China, and yield to the habits of low fertility, Singapore-style. Now, with fewer child deaths but also fewer children, these societies grow older, as we have seen in Catalonia. Eventually, the number of elderly people rises sharply, as in Japan.

Just as with Singaporean fertility rates and the Catalan median age, the number of Japanese centenarians is significant in itself: Japan is a leading economic power with a global influence that encompasses everything from architecture and interior design to food. But it is also significant as an extreme example of a global trend. And although Japan leads the way in terms of the number of super-aged, much of the rest of the world is not far behind.

In the UK, for example, the number of centenarians increased from 4,000 to 15,000 between 1990 and 2015, and it is predicted to reach 200,000 by the end of the century. In 1950, China had just one and a half million people aged over eighty. By 1990 this had risen to seven and a half million people aged over eighty. By the middle of this century, that figure will surpass *one hundred and fifteen million* – over 8 per cent of the country's total population – at which point it will have grown *seventy-five-fold* in a century. As a result, China will be a different country, quite different from any previous point in its long history. In the decades ahead, the Japanese term 'rougai' may be set to go global.

## *The Grey Economy*

Japan can be regarded as a laboratory of the future, testing what happens as a society grows old.[12] Today, 28 per cent of all Japanese people are over sixty-five, by far the highest percentage in the world. Italy is set to reach that point in 2030, Germany in around 2035, China in the middle of the century and the US by 2100, at least according to the UN's predictions. The world has never seen societies like this, and if we want to know what they will look like, Japan is our best guide.

Economics provides an important starting point. Once an economic bright star, Japan came to a halt almost exactly as its working-age population peaked in around 1990.[13] While the sharp and unexpected crash may have been triggered by the beginning of the decline in the workforce, the country's inability to bounce back has certainly had a demographic component: Japan has been suffering from the long drag of gradual population decline.

Thirty years on, the Japanese stock market has never recovered the giddy heights it reached in the late 1980s.[14] In the last thirty years, the country has only exceeded 2 per cent annual GDP growth five times, having only failed to do so twice in the previous thirty years.[15] When we hear economists referring to 'secular stagnation', the long-term slowing of economic growth in the developed world, it is worth noting that Japan has long been the leader of this unfortunate pack – and the fact that it is also the leader of a demographic trend is no coincidence. The lack of economic growth has been accompanied by persistently low inflation; Japan's annual inflation rate has only once exceeded 2 per cent in the last three decades.[16]

It is as if economics as we know it, with its tortuous trade-offs between inflation and unemployment, was based on the assumption of a young and growing population. When this assumption no longer holds, the result seems, at best, to be little growth, which along with persistently low inflation seems to be immune to low interest rates and major fiscal stimuli. Indeed, *without* these stimuli, there would likely be slump and deflation.

In an ageing UK, industrial disputes are running at a fraction of their 1970s and early-1980s level.[17] This is despite more-or-less full employment in many economies, which once seemed to be a guarantee of labour militancy. The global economy, along with the global demography, seems to have become senescent.

The most powerful reason why ageing societies struggle to grow economically is the decline in the size of their workforce. Japan may be an advanced case, but this is also true of countries like the US, which once had a buoyant economy to match its buoyant population.

Think of economic output as the output of each person added

together. More people can produce more goods and services, and the more skilled and educated they are, the more they can each produce. So economic growth springs from both population growth and productivity, which increases with skills, knowledge and education – with both things combined representing 'human capital'. Analysis of the US suggests that the impact of the slowdown in the growth of the workforce since the start of the twenty-first century has exceeded the impact of the increase in their education and experience; human capital is dragging growth towards negative territory. By contrast, in the 1970s and 1980s, growth in human capital contributed to annual GDP growth of more than 1.5 per cent.[18]

An older workforce does bring advantages. Peak productivity and earning power occur later in one's career, meaning that an older workforce is more experienced, even if it is less energetic. It is probably also inherently less confrontational in its demands, which prevents upward pressure on wages and therefore prices. A workforce that has stopped growing finds it easy to gain and retain employment. And full employment, despite the economic theory, no longer seems to turn into assertiveness on the shop floor – older workers are less likely to seek confrontation and to take risks. The gilets jaunes in France may be continuing the national tradition of 'manifestations' and street action, but the post-modern proletariat, insofar as it can be said to exist at all, is not going to bring down the state. An older society also brings distinct opportunities – and challenges – for entrepreneurs and businesses as the tastes and requirements of the population change. Simple things like a larger font on the labels of particular products can add a competitive edge.

The economic impact of ageing began in Japan and is spreading fast; it may be thanks to population changes that interest rates in the West have been so low for so long. The number of young people entering the labour market is dwindling. By 2050, the number of

Italians aged under twenty-five will be barely half what it was in 1980. The cohort of South Koreans in their early twenties peaked about ten years ago and will have halved by 2050. Meanwhile, our economic model is being kept on a life-support system that involves nearly free money. We might have expected interest rates to be forced up by the elderly selling their bonds to finance their retirement, but other forces that depress interest rates have turned out to be stronger.[19]

We have started to hear a lot about 'post-modern monetary theory',[20] according to which the role of the state is no longer to invest only at times of private sector uncertainty but permanently to increase demand to the level required for full employment. The private sector has become so ossified that it is no longer able to drive the economy without government support, for reasons that are at least in part demographic. The trickle of youngsters joining the workforce, the flood of retirees leaving it and the ageing of the population create an environment in which investors and employees look to the security provided by the state, rather than the opportunities offered by the market.

The consequence of zero or even negative interest rates over long periods has been to boost the prices of homes, bonds and stocks and to increase further the wealth of the older people who have accumulated them. An older population tends to look for shorter, safer returns on its investment.[21] Older people are less likely to start new businesses or enterprises; instead of investing in venture capital funds or in the stock market, they look for the security of corporate or government bonds, driving their prices up, and the rates they bear down.

The ready availability of finance makes it cheaper for governments to run significant deficits. It also increases calls on governments to do so, the conservatism that comes with an ageing population resulting in insufficient demand or investment to maintain full employment without state intervention. The Covid-19 crisis has increased these pressures further.

As an older population invests more capital in low-risk projects, its economy slows further. We have seen this in Japan and in Germany, both of which have dwindling reputations as centres for enterprise and are experiencing sluggish growth.

The economists Charles Goodhart and Manoj Pradhan have recently proposed an alternative view: that a waning workforce will be able to demand higher wages, which will trigger a new inflationary spiral. Japan only avoided this, they say, by tapping into a huge increase in labour supply from around 1990, as China and Eastern Europe joined the global workforce. The pair assert that India and Africa, despite their demographic health, will struggle to become the great workshops of the world, as China did, and the deflationary effect of hundreds of millions of new workers will end. Instead, the world will experience labour shortages and workers will demand higher wages, triggering price increases. Once again, it seems clear that demographic decline will result in an economic slowdown but, according to Goodhart and Pradhan, accompanied by inflation rather than deflation.[22] At the time of writing there are certainly signs of rising inflation across the globe; it is too early to say whether these are a temporary effect of the Covid-19 bounce-back or more deeply rooted.

### *A Public Ponzi Scheme?*

Another economic characteristic of Japan that is spreading to other countries is increasing government debt. In Japan, government debt now exceeds 250 per cent of GDP, and the two countries that are closest behind are also ageing profoundly. Both Greece and Italy are 'Blue Zones' of exceptional longevity that also have persistently low fertility rates; government debt is a reflection of the strain that a country's finances face when its population ages. The Covid-19 crisis has resulted in government debt increasing faster than ever before, but the underlying issue existed before anyone had heard of the virus.

The same effect can be seen around the world, though in a less extreme form than in Japan, Greece and Italy. In the UK, it was generally agreed by all parties at the time of the 2019 general election that the public could not endure any more government 'austerity' after a decade of belt-tightening. However, at no time after the 2008 financial crash had the government managed to spend less money than it took in. Instead, all the cutbacks achieved was a reduction in annual borrowing. The stock of government debt may have grown more slowly, but it still grew, and demography was to blame.

An ageing workforce results in slower economic growth, which limits government tax revenue. In the early 1960s, there were nearly five million births in the UK, with this cohort entering the workforce twenty or so years later; in the first five years of the twenty-first century, there were fewer than 3.5 million, and these people will start work soon after 2020. The same pattern can be seen around the world; fewer workers are entering the workplace, reducing the potential for the government to raise funds through taxation.

Meanwhile, the demand on the public purse rises as the population ages. Accustomed to rising living standards and improvement of public services, electorates in the developed world are convinced that the belt-tightening they have experienced since the economic downturn in 2008 has been exceptional, although the public finances suggest otherwise.

So demography has undermined the balance between government spending and revenue. In one form or another, the state has come to bear the cost of healthcare in most of the developed world, and an older population requires increased spending. In real terms, per capita spending on state healthcare in the UK more than tripled in the quarter-century from the early 1990s.[23] This was partly due to the arrival of expensive life-extending and life-enhancing treatments, but it was also the result of the increasing needs of an elderly population. Healthcare spending is rising more quickly than economic growth all over the world, but the gap is greatest in the countries with the oldest populations.[24]

The other area where an ageing population puts pressure on government finances is pensions. For a long time, the UK kept a tight lid on state pension expenditure by limiting increases to the level of inflation. Although the country's pensioners are, on the whole, well-off, this has more to do with decades of successful capital accumulation – and particularly with rising property values – than with government largesse. But in future, the pressure on governments across Europe will be intense, as the number of new workers sinks and the number of retirees grows.[25] Back when Otto von Bismarck introduced a modest old-age pension for seventy-year-old German workers in 1889, life expectancy was not much more than thirty-five.[26] Many workers paid into it but few lived long enough to reap the benefits; the scheme was highly affordable as a result.

The state pension system is essentially a 'Ponzi' scheme: it works if increasing numbers enter the system, but it breaks down when new entrants dry up and there are too few remaining to pay for those exiting. One response is to gradually raise the age of retirement.[27] Over time, the retirement age will rise to sixty-eight in the UK, but protests in France – not least from students who have not even started to pay in yet – have led to the abandonment of an attempt to raise the pension age from sixty-two to sixty-four.[28] Retirement age is one of the few issues on which even the Russian president Vladimir Putin has been forced to retreat.[29]

In part, company pension schemes and private pensions are a solution to the problem, with people encouraged to join their employer's scheme or to make their own provision for old age. Even today the state pension can barely meet the needs of retirees, let alone provide the standard of living that most people aspire to in their 'golden years'. Many countries have tax schemes to encourage this approach, but with falling real interest rates and increasing life expectancies, the pot that must be built up to guarantee a reasonable income in old age is getting ever bigger. The answer for many people is simply to work longer; the idea of a long, leisurely retirement is a fairly recent concept that is already waning. In the EU, the share of people over the age of fifty-five in the workforce rose from 12 per cent to nearly 20 per cent between 2004 and 2019.[30] In the UK, the number of working people over the age of seventy has risen by 135 per cent in a decade.[31]

This fact was brought home to me by a series of encounters with people who were continuing to work well into old age. In 2014 I met the broadcaster, philosopher and former MP Bryan Magee. Already well into his eighties by then, Magee went on to publish his last book in 2018, the year before he died at the age of eighty-nine. A few years later, I had a brief conversation with the artist David Hockney as we stood next to each other in a queue at an airport. At eighty-two, he was planning an exhibition at London's Royal Academy. (He is in good company among artists: Michelangelo was still working when he died three weeks short of his eighty-ninth birthday, as was Picasso at his death, aged ninety-one.) And in the summer of 2019, I went to a concert at the Royal Albert Hall in London. The piano soloist was Emanuel Ax, performing at the tender age of seventy. He and the Royal Concertgebouw Orchestra were conducted by Bernard Haitink, already past his ninetieth birthday and making what turned out to be his final UK appearance.

We can see the same trend in political life. At one stage in the 2020 US presidential election, seventy-three-year-old Donald Trump was set to face one of three septuagenarian Democratic contenders: Joe Biden, Bernie Sanders and Michael Bloomberg. In the end, Biden won; a seventy-four-year-old incumbent and a seventy-eight-year-old challenger contested the subsequent election, which the older man won.

That older people are continuing to work, whether in the arts, politics or elsewhere, is in many ways a positive thing. First, most sixty-five-year-olds are fitter and healthier than they would have been a generation ago. Second, continuing to work is generally healthier than sudden retirement, particularly if working hours can be reduced gradually. Lois Kettner, aged seventy-nine and living in Wisconsin, works at the checkout of her local supermarket and sees herself as typical of many of her generation. 'These are our golden years, but they've got a lot of tarnish on them these days,' she complains.[32] For whether or not working into old age is healthy, the need to keep postponing retirement can cause resentment among those who had been looking forward to giving up work. It's just one of the factors that is causing political lines to be drawn on an increasingly generational basis.

### *The Rise of Intergenerational Politics*

Where politics was once mostly about class, it is increasingly about age. The British general election in 2017 demonstrated how an ageing population can directly impact politics. The campaign was dominated by the question of funding for 'social care' – the day-to-day care of the elderly. In a bold move against its core support, the Conservative Party suggested that, up to a certain threshold, payment should not be covered by the state but be deferred and covered from an individual's estate on death. The policy, dubbed 'the dementia tax', gave rise to a tremendous backlash, which was followed by a humiliating political retreat by Theresa May, the prime minister.

It would previously have been unthinkable for such an issue to take centre stage in a British general election, but now that older people constitute an increasing share of the electorate, age doesn't just have the potential to dominate an election's agenda; it can also determine the outcome. Older people wield political power because they're more numerous and more likely to vote than their grandchildren.[33] It should come as no surprise that in the UK, older homeowners have been insulated from the government's austerity policies, the effects of which have fallen disproportionately on young working families.

Further intergenerational issues have arisen during the Covid-19 pandemic. The risk of death for someone over seventy-five in England is several hundred times higher than for someone aged between fifteen and forty-four. The general response to the disease, and particularly the imposition of a national lockdown, has reflected the priorities of the older middle-aged and elderly.[34]

As we have seen, the burgeoning 'gerontocracy' in the developed world has put a tremendous squeeze on state budgets and on the labour force, with demand for nursing and social care spiralling. Satisfying this need increasingly depends on immigration from less prosperous countries, as richer countries draw on the human resources of poorer, younger populations.

The prospect of delayed retirement may cause resentment among the middle-aged, but the rising cost of providing healthcare and pensions for the elderly is causing rising resentment among the young. However, age has been an increasingly important determinant in voting for a long time. In 1974 the Conservatives had a 37 per cent lead among upper- and middle-class voters, while Labour had a 35 per cent lead among the working classes; by 2017, although the parties gained very similar overall shares of the vote, the class composition of the difference had almost disappeared – instead, what mattered was age. The Conservatives had barely 20 per cent of the votes of twenty-year-olds, while Labour had almost 70 per cent. But Labour had less than 30 per cent of the votes of seventy-year-olds, while the Tories had around 60 per cent.[35] By 2019 the age basis of voting had become even more pronounced. Age was also a strong predictor of how people voted in the Brexit referendum in 2016.[36] The vote to leave the EU and the Tories' electoral victories since 2010 have been underpinned by the size of the supportive elderly cohort and the comparatively small younger cohort. The fact that the elderly are more likely to vote has also helped.[37]

The same effect was visible in the 2016 American presidential election. The Republicans, once the party of the affluent, have become the party of the old, while the Democrats, once the party of the urban working class, are now the party of the young. A recent college graduate living in Manhattan is much more likely to vote Democrat than a male working-class retiree in the Midwest. In 2016, Hillary Clinton received twice as many votes as Donald Trump from voters aged twenty-nine and younger, while Donald Trump had a solid ten-point advantage among the much more numerous voters aged sixty-five or older.[38] In 2020, Trump's support among the young slipped further.[39]

But while patterns of voting are increasingly marked by generational difference, we should not think of the young as always being supportive of 'progressive' causes. In the 2017 presidential run-off in France, almost half the voters aged between eighteen and twenty-four are believed to have backed the far-right candidate Marine Le Pen, compared to barely 20 per cent of those aged over sixty-five.[40]

Sex and race, as well as age, play an important role in UK and US elections, with white men tending to vote for the Conservatives and Republicans, while women and minorities are more likely to support Labour and the Democrats. This has constituted a major shift from the patterns of thirty or forty years ago, when class was the most important determinant. To put it simply, demographic factors – age and ethnicity – are coming to matter more, while economic considerations matter less.

There are good reasons for this. It is easiest to unionize and organize employees in large workplaces, but fewer people now work in big factories and more are self-employed. Second, societies that were once ethnically homogeneous are now more diverse, which has often elicited a backlash indicated by support for right-wing parties among the indigenous working classes. This is a theme to which we will return later.

Age has become more influential in politics precisely because the proliferation of the elderly poses unprecedented challenges in countries where expectations of the state have increased, while its ability to meet them has become more limited. We can expect the continuing growth of the old and the continuing shrinkage of the young to make generational politics even more important in the decades to come.

## *Growing Old Before They Grow Rich*

Although the problems of ballooning state pension and healthcare costs, sclerotic economic growth and a breakdown of the intergenerational compact are doubtless challenging, they might be regarded as 'First World problems' – they trouble countries that can finance their deficits easily and cheaply, either domestically or on the international money markets. If these countries so choose, they can bring workers from poorer parts of the world to do the jobs that the natives are either reluctant or insufficiently numerous to do themselves.

These wealthy economies are able to offer wages and living standards that attract as much labour as they could possibly want, in order to deal with the problems associated with their ageing population. That tends to mean immigration of the poor and young to the countries of the old and rich. But a tendency for wealthy Europeans and North Americans to head in the opposite direction, whether for medical care or for a sunny retirement, is also beginning to take place. For example, there are around seventy thousand Americans and Canadians living in Costa Rica, including many retirees, and many more travel there for the winter.[41]

One looming question is how countries that grow old *before* they grow rich will cope. Rich ageing societies can continue to function while there are young and expanding societies behind them in the development process, but who will care for these societies in turn? Until recently I had a neighbour in London who, up to her death aged one hundred and seven, was looked after by a series of Filipino carers. Who will care for the elderly Filipinos?

The Philippines will not have a major problem in this regard for the foreseeable future – thanks to pro-natal Catholicism, it has an above-replacement fertility rate and will remain fairly young for a long time – but some of its neighbours will not be so lucky. Thailand is ageing remarkably quickly, thanks to a sub-replacement fertility rate and a sharply rising life expectancy. Already over forty, its median age is higher than that of rich countries like Norway or Ireland. By the middle of the century, the over-sixty-fives will account for around a third of Thailand's population, compared to around 13 per cent now and barely 5 per cent as recently as the mid-1990s; Thailand has aged more than four times faster than France.[42]

Thailand is an example of a country where demographic progress has raced ahead of economic progress, leaving its population and economic profile uncoupled. According to the old models of modernity, development and demographic change go hand in hand. By these lights, Thailand should not have a lower fertility rate than France's, a higher median age than Luxembourg's and a life expectancy only a couple of years lower than the USA's.

Whatever Thailand's problems, China's are much greater, given its scale. Although the Chinese are somewhat richer than the Thais, its ageing is comparably speedy. Even if less developed countries were willing to send their surplus populations to support the ageing Chinese, it would require the entire young population of the globe to do the job. As they age, Thailand and China can expect their economic growth to slow down; both have used up their demographic dividends and will have to cope with a contracting rather than an expanding workforce. China has at least been preparing for this eventuality since its imposition of the one-child policy.[43]

Countries like China and Thailand have long revered the elderly, but far from them being the dominant demographic group, they have always represented a fragment of the population. A single statistic illustrates the scale of the problem. In 2000 there were seven workers in Thailand for each retiree; by 2050 there will be just 1.7.[44] Relying on your offspring in old age is not an option when there *are* no offspring. As one retiree puts it: 'Frankly speaking, we have to depend on ourselves for everything. If you are lucky, you might have a family member volunteer to drive you to the hospital.'[45]

Expectations in Thailand of the services the state is supposed to provide are lower than in the West, while ageing Thais also benefit from the fact that a sixty-five-year-old is likely to be fitter and healthier now than ever before. But neither fact can offset the challenge that a country like Thailand will face. Only extraordinary advances in technology will prevent a lot of elderly people, in Thailand and other countries like it, from dying while lonely and neglected. In the absence of people to care for them, the elderly will have to rely on technology.

### *The Technologies of Ageing and Longevity*

Since Japan is both so old and so technologically advanced, it is unsurprising that it is a country where a lot of the developments in social care are emerging. More than a quarter of the Japanese start-ups worth at least £1 billion are related to care of the elderly. Workers in care homes can now receive a signal when incontinent residents require attention, forewarning them of the need for urgent intervention. There are also devices that track vital signs and indicate irregular heartbeats or breathing,[46] while robotic beds that turn into wheelchairs are also being manufactured. The need in Japan is acute, and not just because of the fast-rising elderly population; unlike many other countries, Japan is reluctant to allow immigration. While there is a visa system for foreign care nurses, fewer than twenty applicants qualified in its first year of operation.[47]

Apart from taking care of the body, technology is being used to support the mental wellbeing of the elderly in a society where an increasing number lack relatives or offspring to visit them. Robots and artificial intelligence are increasingly being used to provide solitary old people with the illusion of company and to keep their minds stimulated. As with the physically oriented aides, such Japanese innovations are finding a growing export market. A robot pet designed and made in Japan is already used in several hundred Danish care homes.[48]

A few years ago, I was working for a business that offered a personal alarm service to the elderly. If the wearer fell, they could press a button that triggered a visit or alerted relatives. This is old technology, but there was talk at the time of new versions that would trigger an alert if the curtains were not drawn before a certain time in the morning, and things have progressed a long way since then. Today, cameras can send an alert if there is movement. You can check if your senile mother is trying to leave the house in the middle of the night. There is a great deal of change in this area, and there will be even more in the future.

### *Unequal and Reversible: The Limits of Rising Longevity*

Who lives and dies has always been a matter of individual luck, but there are discernible patterns at a societal level. The most obvious are simply *time* and *place*; someone born in the developed world today is likely to live much longer than both someone born anywhere at all two centuries ago and someone born now in the world's poorest countries. Sex also makes a big difference. Globally, women live five years longer than men on average, but the gap varies. In Russia, it is more than ten years, a difference that is generally put down to the high rate of alcohol abuse and suicide among men. In the Nordic countries, where alcohol use is less gendered and progressive ideology lessens lifestyle differences between men and women, the gap between male and female life expectancy is barely three years. And in countries with decidedly un-progressive values, the gap is narrower still, perhaps because resources are devoted to boys and men over girls and women.

The greatest attention to inequalities in life expectancy has, in recent years, been focused not on comparisons between the sexes or between countries but on class-based differences. In the UK, the gender inequality in life expectancy at birth almost halved between 1980 and 2012. This was likely down to a reduction in smoking, which had been higher among men, as well as improved treatment of cardiovascular problems, which disproportionately affect men. Many more men used to work in dangerous heavy industries, which also took a toll on their lives. However, while men and women in the top 10 per cent by income can expect to live to just under and just over eighty-five, the most deprived women are not making it to eighty and the most deprived men are not living to seventy-five. Over time, the increase in life expectancy in England has slowed, with the worst-off particularly affected, meaning that the gap

Japan Population Structure: 1950

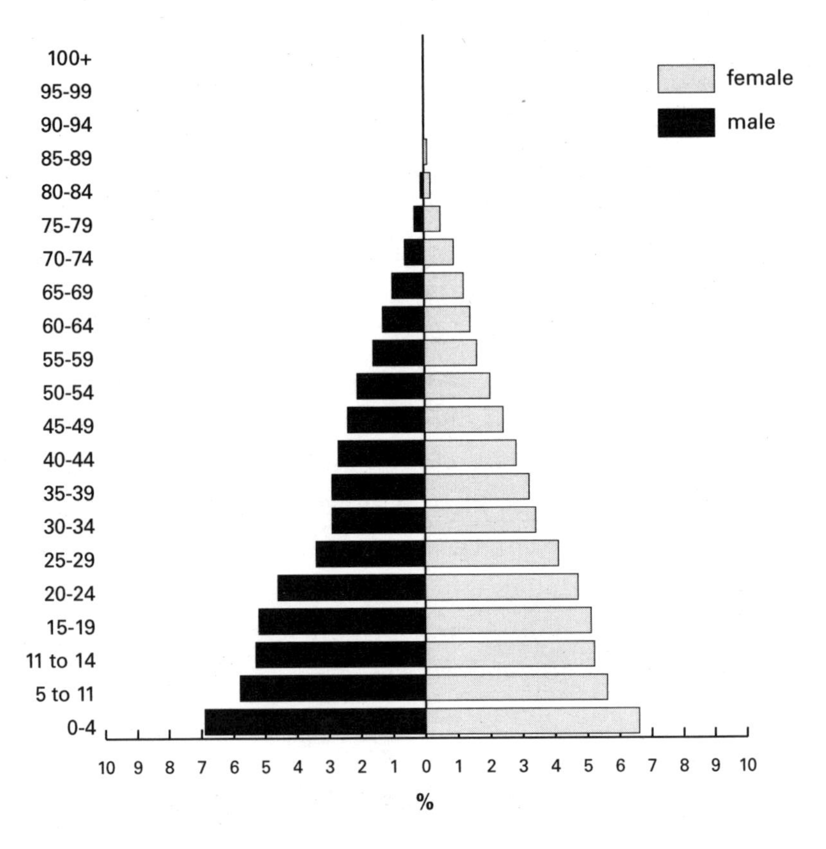


Source: https://www.populationpyramid.net/japan/2019/

Back in 1950, almost every Japanese age cohort was smaller than the one that followed it, a sign of long-term population expansion. For instance, there were almost seven times as many children under five as people in the second half of their sixties. Today, after decades of low fertility and increasing life expectancy, there are more people in their late sixties than under the age of five, and the biggest cohort comprises people in their late forties.

The 'population pyramid' was so-called because it was thought that it was normal for the base to be broad and taper towards the top. However, by 2050, the biggest cohort in Japan will be in their late seventies, and they will outnumber the under-fives by about two to one. The changing shape of Japanese age structure is a vivid illustration of an ageing society.

Japan Population Structure: 2019

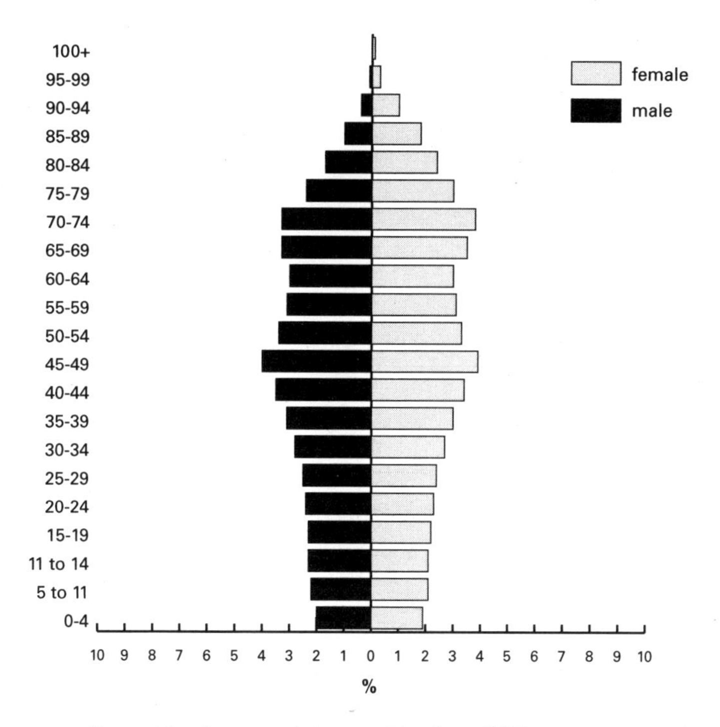


Source: https://www.populationpyramid.net/japan/2019/

Japan Population Structure: 2050

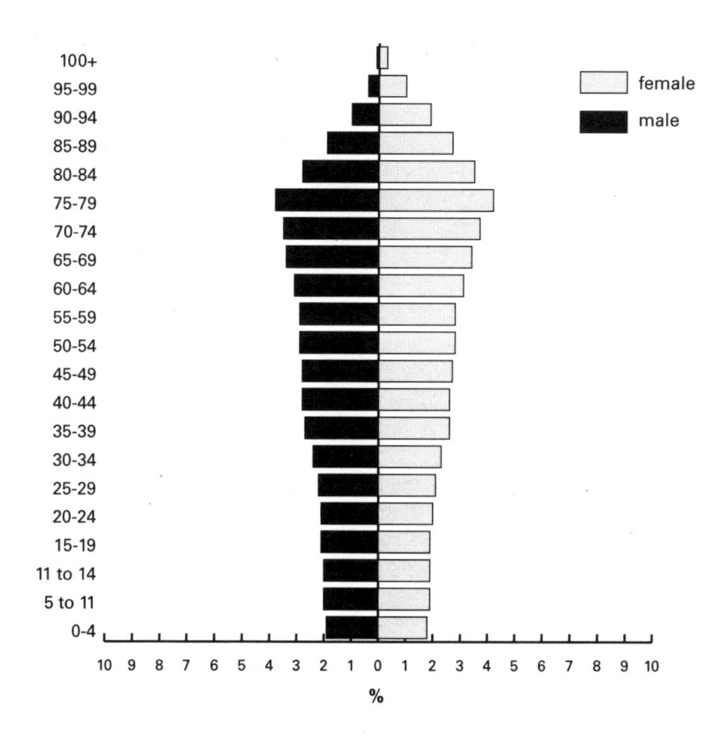


Source: https://www.populationpyramid.net/japan/2019/

between the classes has grown.[49] African, Caribbean and Asian groups in England, according to data compiled between 2011 and 2014, enjoyed a longer life expectancy that whites. (In the category labelled 'Mixed', life expectancy was the same as whites for women and slightly shorter than whites for men.) [50]

Life expectancy at later ages has declined marginally in the UK in recent years. This has a significant impact for life insurers and pension funds, whose liabilities depend on how long people are expected to live. When the Institute of Actuaries confirmed in 2018 that this was 'a trend rather than a blip', several companies were able to bolster their balance sheets significantly.[51] With their members likely to die sooner than they had previously expected, pension schemes could put aside less money for payments.

A similar trend is afoot in the United States, although life expectancy at birth and not just at old age has been falling. As in the UK, there has been much soul-searching as to whether this is part of a continuous trend. So-called diseases of despair, particularly related to drug abuse, seem to be a major part of the explanation, as is the rise of obesity. The relatively easy gains from addressing cardiovascular disease and stroke may have already been reaped.[52]

In the UK many people wish to blame the underfunding of the NHS or the austerity-related decline in public expenditure. There is little doubt that governments could always do more, but the persistence of these trends depends on individual choices. It is clear that a healthy diet combined with a sensible amount of exercise can make a huge difference, but we live in an age in which the state is held responsible for all outcomes.

Despite changes in life expectancy, our societies will continue to age. First, the reversal of the trend is too small and localized to be long-term and general. Furthermore, past projections of increased life expectancy have been too conservative, underestimating just how long people would live.[53] Second, the absence of replacement levels of fertility and the ageing of baby boomers in the developed world is a given. Mortality, however long it is delayed, is inevitable, which means that once societies get old, they will either start to decline economically or need to ship in labour from overseas. And it is to these subjects that we now turn.

# 7

# Population Decline

## *55: Percentage Decline in Bulgaria's Population in a Century*

In 1989, Europe was about to undergo its biggest political change since the Second World War as Communism teetered on the brink of collapse. Meanwhile, in the south-eastern corner of the continent, more than three hundred thousand Bulgarians of Turkish ethnicity were expelled from Bulgaria and forced to flee to Turkey.

Fatma Somersan was a twenty-two-year-old physics student, whose crime in the eyes of the Bulgarian Communist authorities was protesting against enforced assimilation; a few years earlier, over half a million Bulgarian Muslims had been obliged to give up their Turkic or Islamic names and replace them with names that the authorities deemed suitably Slavonic. After a protest against the forced name changes, Fatma was called to the local mayor's office and told to go back to Turkey. 'Now that you have participated in the demonstration,' the mayor said, 'see what it is really like in Turkey'.

Fatma was told to say goodbye to her family and given an hour to pack the one bag she was allowed to take. The Bulgarian authorities, who claimed that they were *allowing* – rather than *forcing* – Bulgarian Muslims to leave the country, laughably named the policy the 'Great Excursion'. But it was a 'holiday' from which there was no return, and for which no compensation has ever been paid.[1] Sadly, this veneer of friendliness that was intended to protect Bulgaria from foreign criticism proved unnecessary; the international community remained untroubled by this transparent act of ethnic cleansing. It was a prelude to its torpor the following decade, when similar events occurred in what was then called Yugoslavia.

The Great Excursion was the most recent chapter in the century-long movement of Christian and Muslim populations throughout the Balkans and Caucasus. Millions of people have been forcibly uprooted – in both directions – between Russia, Greece and the emerging Christian states of south-east Europe on the one hand, and the Ottoman Empire and its successor state, Turkey, on the other. In the nineteenth century, Liberal politician William Gladstone was scandalized by the 'Bulgarian atrocities' of the 1870s, when the Ottoman Turks tried to retain their rule in the face of emerging nationalist and Christian forces. At around the same time, the Russians either wiped out or drove out the bulk of the Muslim populations from places like Chechnya as they extended their rule. Muslims and Christians have persecuted each other in the name of religion in the western Balkans for centuries, and even within recent decades. And more than a million and a half Greeks and Turks were 'exchanged' in the 1920s.

Modern Turkey and Bulgaria were both formed through efforts to establish national and religious homogenization.[2] Where there existed a complex mix of peoples, with various ethnicities and religions, governments tried to create a single predominant ethnic group, with a shared language, religion and sense of common ancestry. For much of the life of communist Bulgaria, which lasted between the late 1940s and the early 1990s, the 'voluntary' emigration of Turks was permitted, if not encouraged, and even, as we have already seen, occasionally enforced. The combination of a socialist and internationalist varnish over an ethno-nationalist core made communist Bulgaria similar to neighbouring Romania, which was at the time ruled by Nicolae Ceauşescu. Here, the emigration of Jews and ethnic Germans was permitted and contraception was easier to get hold of if you were either Roma or ethnic Hungarian. As in Bulgaria, it was an attempt to boost the strength of the majority, while either diminishing minorities or getting rid of them altogether.[3]

When the Bulgarian government indulged in ethnic cleansing in the dying days of its communist regime, the world took little notice. There were no protests on the streets of Western capitals, and the United Nations failed to pass any resolutions. The world was too absorbed by the collapse of Communism, and in any case had little interest in the fate of Balkan minorities. The Bulgarian Turks who left were absorbed into Turkish life, just as generations of other Muslim refugees and hundreds of thousands of Christians fleeing the other way had been before them.

While Bulgaria was getting rid of hundreds of thousands of ethnic Turks, the country's demography was undergoing another significant change. The late 1980s was the point at which Bulgaria's population peaked at just under nine million, up from just over seven million in the middle of the twentieth century. After that it went into decline, and not just because of people like Fatma Somersan being forced to flee.

Today, Bulgaria's population is around seven million, and by 2089 it's likely to have fallen by a further three million. And so, by the centenary of the 'Great Excursion', the country's population will probably have more than halved. The already long-forgotten Turkish expulsions are only part of the story. Bulgaria combines Japanese levels of low fertility with high levels of emigration – these days voluntary – which means that its demographic destiny points towards oblivion. The expulsion of the country's minorities may have been a self-inflicted wound by the government that cut deep into the nation's numbers, but it has been compounded by

Bulgarians' unwillingness to reproduce and by their decision to leave the country when opportunities beckoned elsewhere.

*Population Decline: The Long Term Becomes Now*

Think of a car that is racing uphill. If the driver gradually reduces the pressure on the accelerator, the car will start to slow down. Over time, it will only be progressing at a snail's pace, with less and less thrust driving it forward. Struggling against the force of gravity that is pulling it back, it will finally start to reverse down the hill. This analogy gives a rough idea of how populations go into decline. And for an increasing part of the world, the foot has long been taken off the gas and the car is beginning to slip backwards.

If we ignore migration for the moment, we can think of a country's population size as being determined by two fundamental factors: births and deaths. The forward thrust of extending life expectancy (fewer people dying each year) and high fertility (many people being born) is dissipating, despite the ballooning numbers of the super-old. For much of the world, fertility has been below replacement level for many years, meaning that the forces of 'demographic momentum' have already been used up, while life expectancy gains are slight at best. The engine's thrust is diminishing, and the car is struggling to prevent itself slipping backwards.

The developed world in general, and Europe in particular, is at the tail end of long-term forces that are tending to drag down its population numbers. The growth in life expectancy currently seems to be an almost spent force, at best only modestly increasing overall populations. Long-term low-fertility rates are having a compounding effect, and population decline is the result. Not only are women of childbearing years choosing to have fewer children, but because of the fertility choices of earlier generations, they are themselves increasingly rare. This is how nations disappear.

In Bulgaria, the increase in longevity has been very weak: having already been over seventy in the late 1960s, life expectancy is still less than seventy-five. Fertility rates have been below replacement level since at least 1980, and for much of this period they have been a whole child per woman below where they would need to be to keep the population stable.

The last generation had few children, and these children are now reaching their own fertile years and having few offspring themselves. Think of a woman having a large family – perhaps she has half a dozen daughters. The family's size is growing, but if each daughter has fewer children, growth in the next generation will start to tail off. The death of the grandmother is the loss of one person against quite a few being born. But when the original daughters start to die, if their own granddaughters either decide not to have children or to have one child at most, the balance of deaths and births begins to swing in favour of deaths, and the family size falls.

The total fertility rate in Bulgaria is now 1.5 children, as opposed to 1.25 children twenty years ago, but this is the result of what is known as the tempo effect, the pattern that we discussed in Chapter 4 with reference to Spain. During a period when childbearing is delayed and the average age of mothers increases, there will be a fertility dip; when this stops, fertility will bounce back somewhat.[4] But such minor reversals aside, Bulgaria is typical of the long-standing low European fertility. There are now fewer than half the number of women in their early twenties in Bulgaria than in 1980, so even if each was having the same number of children, the annual number of births to women aged between twenty and twenty-five would have halved.

All this is not just about theoretical numbers or demographers spinning their wheels – it has a real impact on everyday life. Over the Greek border, more than 1,700 schools closed in the five years to 2014, mainly due to a lack of pupils.[5] In neighbouring North

Macedonia, where around a quarter of the population has already been lost, the president has described the country's demographic challenge as its most profound threat.[6] If it joins the EU, and the opportunities for emigration increase, the problem will only get worse.

Germany is another European country that is in poor demographic shape. To keep its population stable, and to maintain the proportion of its population that is of working age, massive population movements will be required. Many Germans who are dying today are from a generation that had few, if any, children and so leave behind no mourners. One indication of this change is that the number of 'public health funerals' – where the state organizes and pays for the funeral – doubled in Hamburg between 2007 and 2017.[7] This trend is probably typical of Germany as a whole.

### *The Third Force: Migration*

In countries like Bulgaria and Greece, long-term low fertility is reducing population size much more quickly than extending life expectancy can maintain it. The third decisive factor, and the one not captured by our car analogy, is the balance of migration. We might think of it as the helpful local mechanic who's prepared to tow the car rather than letting it slide backwards down the hill.

In Germany, a country with more deaths than births, the population would be in decline were it not for immigration. A prosperous country in the heart of Europe, it can offer the prospect of employment and improved living conditions to millions if it chooses to open its borders. We will examine the backlash that can accompany immigration in the next chapter, but whatever the political impact, when Germany opened its doors to those in crisis, as it did during the Syrian civil war in 2015, more than a million refugees moved there.[8]

Germany has long attracted immigrants. In the decades following the Second World War, hundreds of thousands of people moved there from southern Europe, Turkey and the Balkans. Along with immigration from the former Soviet Union, all this helped to stave off population decline. Many came as *Gastarbeiter*, or 'guest workers', and ended up staying. If you walk around just about any German city, at least outside the former East Germany, you'll find the kebab shops, the Turkish barbers and the mosques that are clear signs of generations of immigration. Whatever its advantages and disadvantages, immigration serves as a countermeasure against falling population. A rich country like Germany, whose people are increasingly reluctant to have children, can afford to effectively outsource the business of childbearing and rearing to those in poorer countries, and then attract the children from those poorer countries when they are old enough to work. So when gravity is threatening to drag the car down the hill, immigration can be brought into play. Hundreds of thousands of people will risk their lives to escape from countries like Chad, Afghanistan and Syria to get to countries like Germany.

For a country like Bulgaria, the balance of migration has the opposite effect; rather than offsetting demographic decline, it compounds it, adding to the force dragging the car backwards. Indeed, emigration is responsible for around two-thirds of the population decline in Bulgaria, making it an even bigger factor than low fertility.[9] It's no longer ethnic Turks who are leaving the country, forced out by an oppressive regime; instead, with democracy installed, young and educated Bulgarians are departing voluntarily, attracted by the higher wages and standards of living offered in the countries of Western Europe, to which Bulgarians can freely migrate.

There is some inflow: during the Syrian refugee crisis, Bulgaria was the point of arrival for many migrants coming via Turkey, and in 2014 and 2015, around five thousand people were granted refugee status each year. Most of them would rather have moved on to Germany, but for those allowed to stay in Bulgaria and unable to continue west, it was better than going back to their war-torn homelands. Since then, the border has hardened and the inflows are small compared with the outflows; in 2017, there were almost nine times as many Bulgarians living abroad as there were foreigners living in Bulgaria.[10]

Efforts are underway to reduce the flow of Bulgarians out of the country and to attract people back, but the impact so far has been modest. As one returnee lamented in 2019: 'We have a whole Bulgarian capital living abroad.'[11] People of working age generally don't tend to migrate from high-income countries to lower-income ones, and they rarely want to go back to low-income countries once they have lived somewhere more affluent. And most Bulgarian expats are in rich countries. The onset of the Covid-19 pandemic in March 2020 caused around 200,000 Bulgarians to return to their country,[12] but it will struggle to keep them once life returns to normal.

### *The Emptying Countryside*

When the population explosion takes place in a predominantly rural country, whether nineteenth-century Britain or modern-day Nigeria, there is a vast overspill into the towns, which grow like mushrooms. But despite the movement from the villages to the towns, the countryside does not initially empty out, but rather loses its excess population that cannot be absorbed into farming. Long after the population explosion is over, however, cities continue to entice country-dwellers, with their bright lights, well-paid jobs and opportunities for an exciting life. And since those who remain in the rural areas are no longer having lots of children, the overall population falls. Villages become hamlets, which eventually become clusters of just a few houses. Eventually, all that is left are crumbling ruins and a lone farmhouse, which will itself ultimately be abandoned.

Massive population loss in Europe is nothing new: the Black Death, a continent-wide catastrophe in the fourteenth century, caused Europe's population to decline by almost a third.[13] And two hundred years later, parts of Europe saw their population drop by a similar proportion during the Thirty Years War.[14] China has seen significant falls in population throughout its long history, whether from floods or plagues. When Europeans arrived in the Americas, local populations collapsed.[15] What is different with the current population decline in rural areas around the world is that numbers are not falling as a result of some horrible external force like war or disease; instead, it is purely the result of choices made by men and women about how many children to have and where to live.

Throughout history, most populations have been fragile, making tentative gains that are later eroded. We can see this represented in Leonard Woolf's 1913 novel *The Village in the Jungle*, which was inspired by the author's time working as a colonial administrator in Ceylon (now Sri Lanka). The villagers in the novel suffered a perennial struggle to survive. At the end of the story, despite their best efforts, the jungle begins to creep back and the village is gradually effaced; nature, the message seems to be, triumphs wherever people retreat. A bad crop or an outbreak of disease could wipe out years of population growth and prompt a decline from which a community might never recover. A village could come into being as an overflow from a larger settlement, before being knocked into oblivion after just a few generations.

Today's retreat is occurring across a vast swathe of the world where fertility rates have long been low, but sticking to the example of Bulgaria provides a good picture of what is happening. A survey revealed that between the middle of the twentieth century and 2012, the population of the Bulgarian countryside declined by around 60 per cent, a trend that has continued in the decade or so since then.[16] The number of settlements that are big enough to qualify as villages fell from around six thousand to around five thousand between the end of the Second World War and 2007, and things have got even worse since then. Unsurprisingly as a result of the 'Great Excursion', villages that were once populated by ethnic Turks have seen the greatest fall in numbers.[17] While Bulgaria's villages are being depopulated at an alarming rate, the elderly tend to be left behind to nurse memories of a time when the valleys echoed with the sounds of children playing. They might be left wondering where a priest will be found to administer them the last rites. 'All my friends that I grew up with here left long ago,' one of the few thirty-somethings who remained in a village not far from Sofia told a BBC journalist. The shop in his village is sparsely stocked – its owner does not know when she might have to shut for lack of customers. And in villages further up the valley, the shops have already closed down.[18]

In Vidin, a province in the remote north-west of Bulgaria, tucked between Romania and Serbia, things are even worse. The region's working-age population has halved since the 1980s, causing the area to enter a spiral of decline. It was once served by a thirty-minute domestic flight from the capital, but Vidin can now only be reached by a tortuous five-hour car journey. One local from the provincial capital, who has previously tried to leave before having to return each time the work ran out, complained: 'It was as if I were coming back to my grave. This is a dying city.'[19]

The process of rural depopulation becomes self-reinforcing. Once the population of a village falls below a certain level, it can easily lose its school, meaning both that young families will not be attracted to live there and also that people who are already there might choose to live elsewhere. Basic amenities such as the bus service, the bakery and the grocery shop tend to shut down once a critical mass of population is lost; the costs of public services are no longer justifiable, while local businesses become unviable. The regional capital, surrounded by declining villages and towns, attracts less investment and loses its transport infrastructure, making the area increasingly isolated and less attractive as a place to live.

Of the many countries across central and eastern Europe that are experiencing radical population loss, Russia is a particularly striking example. As in Bulgaria, long-term urbanization has been compounded by a historically low fertility rate, dating back at least to the 1970s. The result is a country that has more recorded deaths than births. Unlike Bulgaria, Russia has experienced significant immigration, especially from the countries of the former Soviet Union, while emigration is limited by its not being in the EU. However, immigrants to Russia, just as in most places, tend to be attracted to the bright lights of the big cities rather than to the dying countryside. Twenty thousand Russian villages have been completely abandoned and another thirty-six thousand have fewer than ten inhabitants.[20]

The harshness of the climate in much of Russia and the remoteness of so many places in this vast landscape are both factors that have contributed to the decline of the country's provinces. Russians have retreated to bigger urban centres, where homes are centrally heated and a variety of groceries can be more easily and affordably purchased. The sad tales of the elderly remaining in once-lively villages that are now dying are pretty much the same as in Bulgaria, as well as in a growing number of places across the world. 'There are just a few kids left in school. How long will the old people survive?' asked Vera Selivanova, a social worker from Shelepovo, not far from the border with Kazakhstan. 'The village is dying and no one cares,' she told a visiting reporter.[21]

### *The Geopolitics of Empty Spaces: Siberia, the Russian Far East and China*

Rural depopulation in Bulgaria has few implications for international relations, but the same cannot be said about the Russian abandonment of vast tracts of their country. The Russian state has for centuries been nervous of the sparseness with which much of its territory is populated and has long made efforts to turn the tide.

Alaska was part of Russia before it was sold to the USA in 1867, and Russian settlers travelled down the Pacific coast and as far as northern California in the early nineteenth century. In addition to the presence of Russian settlers in North America, the extension of railways into Siberia consolidated Russia's hold on the area from the 1890s. Stalin's industrialization of the Urals in the 1930s, when men and women were moved to the far-flung frontier, along with machinery to encourage development, was a landmark development in Russia's outward movement of its population. It also played a critical role in the victory over Nazi Germany, by ensuring that not all industrial areas were overrun. And in the 1950s, Khrushchev's 'Virgin Lands' campaign, through which young pioneers were encouraged to settle on the outer edges of the country, was the last gasp of Russian expansion.

Since then, however, there has been more than half a century of retreat. Some of this has been political, caused by the demise of the Soviet Union and the breaking away of all the non-Russian republics, but there has also been demographic decline. The Russian presence in the remote corners of its territory is coming undone as populations age, having failed to reproduce – and when they do have children, they find that their offspring don't share their desire to live in the back of beyond. President Vladimir Putin has expressed worry about the Russian Far East, a so-called red zone with a declining number of Russians that abuts China, whose population is still growing.[22] Putin's recent efforts to offer land to would-be farmers in an attempt to resettle the east are having little effect; the soil is often too poor for agricultural use, while bureaucratic obstacles tend to slow down the process to a grinding pace.[23]

The lack of people to manage the infrastructure in the Russian Far East is making the region less habitable. Businesses struggle to relocate relevant experts to the region, while most locals who have qualifications in economically useful fields such as oil exploration prefer to work in the big cities of the Russian west.

Although Russians may cast an anxious eye over the Chinese border, worrying about their neighbour's large population in adjoining provinces to those that Russians are deserting, many villages in China are also dying. This is, in part, caused by the massive urbanization that we have already charted, but if the Chinese were still having four or five children, or even two or three, there would be plenty of people left for the villages. After all, the Nigerian countryside is not emptying out, even as its cities are filling up.

Lumancha, a village in the north-western province of Gansu where there are almost no adults under the age of forty, is a typical case of Chinese depopulation. 'In the past, when it got warmer after the winter, there would be many children running about and playing, shouting, having fun,' the headmaster of the village's primary school told a visitor. 'But nowadays, even when it's the school holidays, you hardly see a child anywhere. Not during the summer holidays, not during the winter holidays. The children who go to the cities to study don't come back.' Just a decade ago, his school taught a hundred pupils; now it has just three, and it is one of nearly two thousand schools in Gansu alone that have fewer than ten pupils.[24]

China's problem is not just a localized one. Persistent low fertility over the past four decades, intensified by the one-child policy but not reversed by its lifting, has meant that the contraction of the Chinese population is on the horizon. During the past decade, annual population growth in China has been at around 0.5 per cent.[25] Towards the end of the decade the growth rate has been slower and some believe China's population is already falling.[26] China is at least on the cusp of population decline. When India's population size overtakes China's, almost inevitable in the next few years, it will be the first time since China became a state – that is, the first time in more than two thousand years – that China has not been the world's most populous country.

### *Depopulation Comes to Town*

Japan, as we have already seen, is another society that has experienced a persistently low fertility rate. As a result, it is not only ageing but emptying out. Japanese villages, just like those in Bulgaria and Russia, are being abandoned. Wildlife flourishes in countryside that has been abandoned by farmers, before beginning to encroach on dwindling human habitations; in northern Japan, bear sightings doubled in a single year.[27] But now even the country's suburbs are starting to empty out. In much of the West, many middle-aged people look forward to the financial windfall of inheriting the home of an older relative. In Japan, by contrast, in order to avoid the high rate of taxes on second homes, many people who inherit properties choose to renounce their ownership and declare them abandoned. Nearly one in seven homes in Japan is registered as ownerless, and the problem is set to get even worse.[28]

Increasing numbers of tourists may travel to see the bustling centre of Tokyo, but just a short train ride away from the capital, its suburbs are both ageing and emptying. While the problem is worse in provincial towns, one real-estate expert has predicted that in fifty years' time, Tokyo's suburbs may be like mini-Detroits, with a proliferation of damaged, empty properties.[29]

This reference to Detroit reminds us of what has been happening in much of Europe and North America. Although a country's population may be stable or even modestly growing, as a result of immigration and the residual weak effects of demographic momentum, the ceaseless outflow of people that began in the countryside has now spread to some towns. The worst affected are cities like Detroit, which invested heavily in industries that have since gone into decline and have consequently become urban rust buckets with diminishing employment prospects, declining commercial centres and rotting infrastructure.

The death of towns is a symptom of a sickness that exists across much of the developed world. A friend of mine who was brought up in the once-thriving English pottery town of Stoke-on-Trent recently commented that on a recent return visit, it was unrecognizable from the place of his childhood. Although many people had been poor when he was growing up there in the 1930s and 1940s, the shops and streets had been thriving. The present-day inhabitants of Stoke are better fed, housed and educated, and unemployment was quite low – at least before the Covid-19 pandemic. They own smartphones and enjoy cheap overseas holidays that would have seemed miraculous to my friend in his youth. But despite all this, my friend felt a sense of despair about the place.

The elderly tend to see things through rose-tinted spectacles, so some scepticism may be in order here. Demography cannot fully explain urban decline – in fact, it's as much part of the effect as it is the cause – but the population data can provide an indication of the sense of decline my friend felt. Stoke's population, which grew fifteen-fold between the start of the nineteenth century and the 1920s, peaked in the middle of the twentieth century and then went into a gradual and erratic slump. But more striking is the change in its age structure. Before the First World War, there were at least four times as many children under the age of five than there were people over the age of sixty-five; now, the latter outnumber the former by around two to one.[30] Furthermore, the town has lost 40 per cent of its bars and pubs in the last twenty years.[31] It's not surprising that Stoke feels like a different place now from the young and busy city my friend knew seventy years ago.

The lack of people, and particularly the lack of young people, is a central theme in the story of urban decline. If Britain's fertility rate had held up and its population had maintained the growth it experienced for much of the nineteenth and early twentieth centuries, places like Stoke would not now feel so reduced. There would be enough people to go round in order that more places could experience the kind of population growth that has been achieved by successful English cities like Manchester and Liverpool, both of which have reversed population decline in recent years. It should, however, be said that the success of these northern cities has arguably been at the expense of surrounding towns, which feel annexed or hollowed out.

The difference between population growth and decline is something that you can feel tangibly as you walk through a city centre. Visit the English university city of Cambridge, for example, and it is clearly thriving. There are tell-tale signs: the presence of people in the city centre even out of term time, while most of the commercial outlets are occupied and functioning. Restaurants and bars are full, and there are no boarded-up high streets. Having seen this, it's no surprise to discover that Cambridge's population has grown steadily over a long period. It roughly doubled between the 1920s and the census of 2011, while Stoke's population fell by around 6 per cent during the same period.[32] Cambridge presents a very different picture to Stoke, and what you can feel on the streets, you can also see in the population numbers.

Another English city, Sheffield, has maintained its population through an increase in the number of students, now over 60,000, despite a decline in the number of manufacturing jobs since the early 1970s from 125,000 to just 25,000.[33] It may be more pleasant to be a student than a steelworker and an educated population does bring benefits, but while the steelworks at least at one time paid their own way, the students are funded by a rising tide of debt.

What my friend felt about Stoke applies across much of the developed world. The United States has a vast 'Rust Belt', where industry has moved away and left behind towns with declining populations. Streets that were once busy are now empty and disfigured by boarded-up shops. The same problem can be seen in wide tracts of Germany and in France. It is true that the populations of France, Britain and the United States are not yet declining in absolute terms, but our urban geography, like our economy, is built on the assumption of ongoing, vigorous growth. Once that ends, cities seem sapped of their energy and dynamism.

It is no longer just the smaller towns or those in the Rust Belt that are experiencing population decline. Even the centre of Paris has seen a decade of decline; fifteen schools closed or merged in the three years to 2018. A hundred thousand more Brits leave London than arrive each year, a number offset only by mass immigration. New York has also recently started to see its population shrink – and all this was before the Covid-19 pandemic.

The great population boom of times past began in the countryside before flowing into the towns, and this continues to be the case in Africa. Similarly, population decline is first seen in the remotest hamlets before working its way inward, until it can eventually be seen in abandoned apartments in Tokyo suburbs and closed-down Parisian bakeries. This is what long-term population decline looks like.[34]

## Selected Countries' Population, % of 2020 Population, 1950–2100

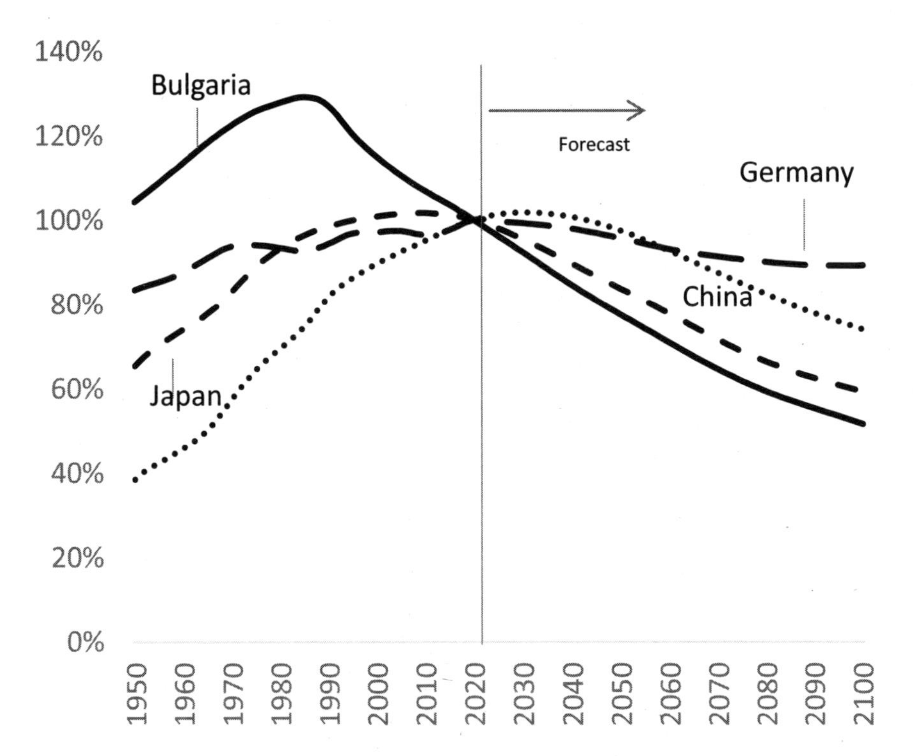


Source: United Nations Population Division (Medium Variant Forecast

Historically, populations normally only contracted if they were hit by famine, plague, war or other calamity. However, decades of low fertility have left countries facing self-induced population decline. Bulgaria experienced this earlier than others, hastened by emigration, but Japan is there now, while Germany is only able to avoid population decline thanks to net immigration. China's population has already passed peak working age, and its overall numbers will soon start to fall.

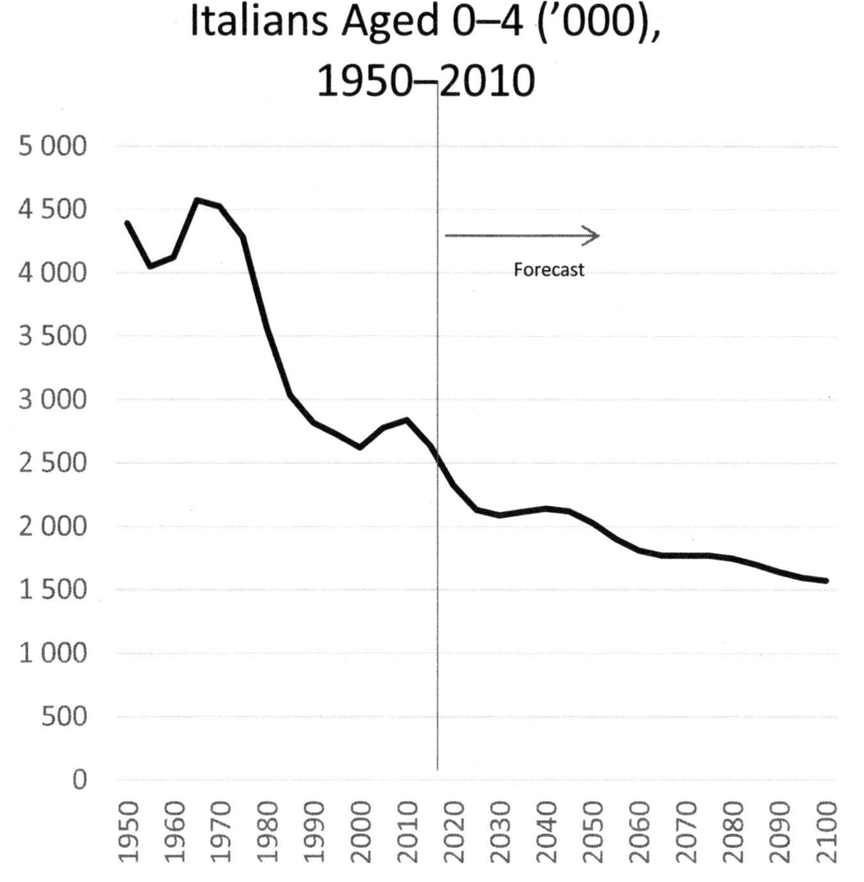


Source: United Nations Population Division (Medium Variant Forecast)

Population decline can be seen most starkly in the falling numbers of the young, and the situation in Italy is a case in point. By the end of this century, the number of Italian children under the age of five is forecast to be barely a third of their peak, which was sixty years ago.

### *Tomorrow's People: Will There Be Any?*

The above description makes depopulation seem like an inevitability, a creeping disease that starts in remote rural areas, the body's extremities, before gradually progressing all the way to the heart. Just like a virus, it has spread from country to country, continent to continent, cropping up in places that were once associated with bustling crowds and enormous families. The stories that once seemed localized, of deserted hamlets, shuttered schools and empty suburban flats, seem to be going global. Where once it looked like people would consume the planet, it now looks as if we might eventually be asking the last human left to turn out the lights.

A 2019 book called *Empty Planet* by Darrell Bricker and John Ibbitson summed up this thesis very effectively. The whole planet is urbanizing, while also introducing improved rights for women and more liberal attitudes in general. These things will mean that fertility, already low, will drop further and low rates will spread wider, eventually leading to population decline. While plenty of data is pointing in that direction, Bricker and Ibbitson think it generally understates the case. In India, they write, 'over and over again . . . demographers and government officials . . . told us, *sotto voce*, that they suspected the fertility rate had already dropped below 2.1.'[35] Latest data shows they were right. Equally suspicious of the African data, the authors think global population will peak and begin to drop not at the end of the current century but in its middle, in just a few decades' time.

The panic seems to flipflop over the centuries. In the early nineteenth century, when the demographic take-off in England was getting underway, Thomas Malthus predicted a future where too many people would overwhelm the ability of the planet to sustain them. A hundred years later, when people started to notice that family sizes in England were beginning to fall, the *Daily Mail* started to fret about 'the decline of the race' and US president Theodore Roosevelt spoke of 'race suicide'.[36] By the 1960s, when global population growth was at its peak, the biologist and population expert Paul Ehrlich threatened us with his 'population bomb', having looked out of a taxi in Delhi to see 'People eating, people watching, people sleeping. People visiting, arguing and screaming [ . . .] People defecating and urinating. People clinging to buses. People herding animals. People, people, people.'[37] Now, however, we seem to think we will soon be wondering where all the people have gone.

Both sides in this centuries-spanning debate tend to be too keen to see current trends as irreversible. To get a more balanced picture, we need to return to the question of fertility, which will ultimately be the determining factor. Low fertility is, after all, at the root of population decline, and the empty planet will only be avoided if low fertility is reversed, or at the very least, if it fails to spread everywhere. Urban populations need not necessarily have low fertility rates. It is true that Kolkata has a fertility rate of little more than one child per woman. But in Lagos State, which is dominated by one of the world's largest and fastest-growing metropolises, women are still having around twice as many children as the replacement rate.[38] That may, of course, tumble, but it may not. It is not some impersonal social force that will determine this, but the life choices of millions of women and men.

Just as cities that are at a similar level of development can vary enormously in their fertility, the same can be true of countries. There are cases like Thailand, where fertility has fallen from 5 to 1.5 since the 1970s, but there are also cases like Sri Lanka, where fertility has been somewhere between 2 and 2.5 for the best part of three decades.

Even within the developed world, there are differences between countries that are worth drawing out. Japan and the countries of southern and eastern Europe may show little enthusiasm for childbearing, but the fertility rate of the Nordic countries is close enough to replacement level that any natural population decline is very slow; the rate in Denmark and Sweden has remained more or less consistent for fifty years, while Sweden's fertility rate was lower in 1937 than it is today.[39] The point is that it is possible for fertility rates to remain at or near replacement level for generations; a drift to sub-replacement fertility is by no means inevitable.

Although we often speak of trends at a continental level, there are significant differences *within* continents. Fertility tends to be higher in west Africa than east and southern Africa, and much lower in north Africa; southern and eastern Europe have rock-bottom rates, but northern or western Europe have not – at least yet.

The deeper fall in East Asia than South Asia can be attributed to greater advances in material conditions – China and Japan are economically well ahead of India and Pakistan. But this straightforward relationship between development and low fertility does not hold in Europe, where affluent Scandinavia, France and the UK out-breed Italy, Spain, Greece and – of course – Bulgaria. Nor do levels of development seem to explain why Nigerian women are slower to reduce their family sizes than women in Kenya. The old explanations of modernization only get us so far; more interesting stories of culture, tradition and belief are needed to give us a more comprehensive picture. Post-modern demography is creeping in.

Whatever the variations in fertility within and between continents, the global population will not be falling any time soon. If UN data is to be believed, the global rate is still well above two children per woman. Even when it does fall below replacement level, 'demographic momentum' means that populations will continue to grow because there will be so many young people to have children, thanks to the large family sizes of previous generations. In addition, in many countries life expectancy will continue to grow significantly – the car driving uphill still has a lot of forward momentum. For similar reasons, European populations kept growing through the second decade of the twentieth century, despite the devastation of war and the Spanish flu pandemic.

In the wake of Covid-19, we are far more aware of the risk a pandemic could pose to the global population – the Black Death set back some European populations for centuries. Some people have speculated that the reason we have not yet found intelligent life on other planets is because something or other will always kill it off – whether that is its own self-destructiveness, longstanding low fertility or some kind of virus or germ.[40] However, for the time being it is safe to assume that humans will be around for a while and that 'peak person', never mind 'zero person', is still some decades away.

In much of the rich world, but particularly in Western Europe, population would be falling were it not for immigration. In Germany, two hundred thousand more people die each year than are born. Manfred Grosser, a clergyman in a town between Berlin and Dresden, officiates at five funerals for every baptism and talks of there being 'dark demographic clouds on the horizon'.[41] Without high levels of immigration, Germany's population would soon be falling, and by the middle of this century it would be losing half a million people every year. By the end of the twenty-first century, Germany's population looks set to be 40 per cent below its current level. Indeed, without relatively high levels of fertility among immigrants to Germany, the situation would be even more dire.

Western Europe and North America may be able to stave off population decline through attracting immigrants, who come from Latin America, Africa, the Middle East and Asia. This is transforming the United States and Canada, and it is causing the most rapid change in the ethnic composition of Europe since the movement of peoples that followed the collapse of the Roman Empire. And it is to the subject of this ethnic change that we now turn.

# 8

# Ethnic Change

## *22: Percentage of Californian Schoolchildren who are White*[1]

'The United States . . . was God-intended, I believe, to be the home of a great people. English-speaking – a white race with great ideals, the Christian religion, one race, one country and one destiny. It was a mighty land settled by northern Europe from the United Kingdom, Norsemen and the Saxon . . . The African, the Orientals, the Mongolians, and all the yellow races of Europe, Asia and Africa should never have been allowed to people this great land.'

So thundered the congressman Ira Hersey in a 1924 debate over the introduction of a bill to control immigration into the United States. The US had in the years prior to the First World War received a vast wave of immigrants from southern and eastern Europe. Only once that conflict was over, in a spirit of isolation, were the first serious limits on immigration from Europe introduced. These controls were unashamedly aimed at preserving the ethnic character of the country; whether that was defined as 'Anglo-Saxon' or 'northern European', people from the British Isles were favoured over those from Russia, Poland or Italy. Quotas were based on the ethnic mix of immigrants in the US population in 1890. The restrictions of the 1920s were aimed at limiting the 'wrong kind of European', but also banned immigration from Asia entirely. The congressman Albert Johnson, one of the bill's sponsors, made his goals clear: 'Our hope is a homogeneous nation . . . Self-preservation demands it.'[2]

But Albert Johnson's outspokenness was modest compared with that of his supporter Ira Hersey, whose bigotry recalled Boss Finley, a character in *Sweet Bird of Youth* by Tennessee Williams. Despite his racially charged and religious rhetoric, Hersey was a representative not of a Bible Belt or southern state but of the northern and overwhelmingly white state of Maine – bigotry was clearly widespread in America in the interwar period. Hersey was clearly confident about God's preference for people of north European origin, which indicated an ethnic hubris that was only possible at a time when people of European origin had the world at their feet and expected it to stay there.

### *Immigration and Race in America*

The United States has long wrestled with issues of immigration and race. It has always desired more people and development, while also struggling over who would make a suitable American. It has also long been inhabited by some people who are eager to embrace humans regardless of their origin, but these liberals have continuously rubbed up against others with a more ethno-national outlook.[3]

In the nineteenth century the United States was driven by an impulse to expand, which meant filling its empty spaces with people, cities, railroads, factories and farms. This sense of mission, which became known as 'manifest destiny', was about the idea that Americans were destined to build a great nation from coast to coast. It was an impulse that was messianic, ideological and practical. Many Americans thought of themselves as being called on by God to spread into the wilderness, but there was also a strong economic impulse. To fill America, people were required, and in large numbers. Americans were fertile, with large families and a relatively high survival rate, but the country's rapidly growing population was not enough to fill a continent with the urgency that manifest destiny required. And so it was that the US accepted – and even welcomed – the poor and unwashed from Europe's furthest corners. They even erected the Statue of Liberty to make the point.

The people who arrived at Ellis Island in the decades prior to the First World War were different from the English, Scottish, Welsh, Irish, Dutch and German immigrants who had preceded them. By the late nineteenth century, thanks to improved transport both within and from Europe, the journey across the Atlantic had become a feasible prospect for people from places like Sicily and Poland, for whom America had previously felt like a very remote place. But by the turn of the twentieth century, America had become a place to which many more people from Europe's interior could aspire to migrate, and the process gained momentum. As we saw with migration from Africa in Chapter 2, once an uncle or cousin has settled, they are a familiar face to meet the next arrival, with a bed for the first night or two and a few useful connections to help find work.

America's desire to grow and to build delayed its imposition of immigration controls, but its approach before the 1920s was far from completely liberal. In 1848, when California and much of the American West was annexed from Mexico, the Mexicans who lived in these conquered lands were not regarded as desirable citizens of the republic. Territories were only recognized as states of the Union when they had an established white majority – until then, they were administered territories. California had a significant Mexican population but it was highly attractive to incoming Americans and so quickly gained a white majority, being accepted into the Union in 1850. New Mexico had a bigger Mexican population to start with and was less attractive to white settlers; as a result, it did not become a state until 1912.[4]

Sentiments about ethnicity impacted not only *when* territories were admitted as states but also *which* territories were included. When it was suggested that the Philippines be annexed after the Spanish War of 1898, Senator Ben Tillman of South Carolina protested: 'You are undertaking to annex and make a component part of this government islands inhabited by ten millions of the colored race, one half or more of whom are barbarians of the lowest type.' The effect, he said, would be to inject 'into the body politic of the United States that vitiated blood, that debased and ignorant people'.[5]

California's economic prospects, and particularly its gold, drew in not only Europeans but also Asian immigrants in the decades following the conquest of the west by the US, which gave rise to a strong reaction. As early as 1852, taxes were imposed on the Chinese inhabitants of California, with the goal of discouraging their settlement. In the late nineteenth century, various acts were passed to restrict their presence, and these were often accompanied by violence against them. Trade unions often favoured immigration controls because unchecked foreign labour meant competition and lower wages; businessmen tended to favour immigration for the same reason.

Once California was part of the US, it was rapidly populated by a tide of Americans of European extraction. Over the course of the twentieth century, the state's population rose from less than one and a half million to over thirty million.[6] The new arrivals were attracted by the state's rich golden farmland, as well as driven to flee the poverty they were experiencing back east. This push and pull process was still going in the 1930s; in John Steinbeck's 1939 novel *The Grapes of Wrath*, the unfortunate Joad family are driven from Oklahoma by the emergence of the Dust Bowl and travel, as thousands did, to the Pacific Coast, which was still regarded as a promised land.

However, by the end of the twentieth century, the new Californians were coming from the south rather than the east, in a great

Hispanic wave. The non-Hispanic 'European' population is now in a minority, and in schools it is an ever-smaller one. Two generations ago, whites were the overwhelming majority in Californian schools. One generation ago, over 40 per cent of California's school population were white; today, the number is 22 per cent and falling.

### *The West Becomes the North*

For the United States, the annexations that followed the war with Mexico in the mid-nineteenth century represented a great opportunity for westward expansion. The nation had been moving in that direction ever since its founders arrived in New England and Virginia. But from a Mexican perspective, this area is not the west but the north, and a vast territory that was lost.

The speed of demographic change in the American south-west in recent decades has been staggering, as the figures cited above illustrate – and the school population can be taken as an indication of the future overall population. In 1970, over 75 per cent of California's population were white and 12 per cent Latino. By 2018, 38 per cent of the population were Latino, while 37 per cent were white.[7] In recent years, more Mexicans have been leaving the United States than have been arriving, but immigration from Honduras, Guatemala and El Salvador has increased. The Californian population of tomorrow can be seen in its schools, where Latino pupils outnumber their white peers by more than two to one.[8]

This is all the result of the shifts we have identified in previous chapters: the long-standing low rate of childbearing in the developed world, coupled with the high fertility and rising child survival rates in the Global South. Today, the fertility rate in Mexico is not much higher than that of the US, but during the 1970s it was three times as high.

While demography created the conditions that enabled mass

population movement, economics provided the catalyst. A dynamic US economy has demanded cheap labour in the last four or five decades, just as it did prior to the First World War. But now, unlike then, Europe is wealthy and not very demographically fertile, which means its population has neither the demographic heft nor the economic motivation to migrate to the US. And for the populations of less prosperous Eastern Europe, emigration to Western Europe is a more straightforward option.

So for Europeans, America is no longer the destination for population overspill that it was a century ago. Instead, cheap labour is now coming to the US from south of the Rio Grande. In the first decade of the twentieth century the sweatshops of New York's Lower East Side teemed with the population surplus of Russia, Italy and the remoter corners of the Austro-Hungarian Empire; by the end of the twentieth century, the gardens and pools of wealthy Californians were tended by migrants from Mexico and Central America.

Poor and until recently demographically fertile, Latinos are responding as once Europeans did, and it's not just in California that the makeup of the US population has been affected. The white and Latino populations in Texas were similar in 2019.[9] As recently as the 1980s, two-thirds of the Texan population was white; that figure is now barely 40 per cent, and in two decades' time it is expected to be less than a third.[10] At a national level, the Latino population is already significantly bigger than the black population. The 2020 census showed that less than 60 per cent of the American population self-identify as white. It is forecast that by 2060, less than half of all Americans will be white, and the Latino population will be more than twice the size of the black population.[11]

Such predictions make matters of race seem cut and dried, whereas they are sensitive and complex. For a start, the data is based on how people self-identify, which is subjective and can change. 'White' was a less meaningful category when American elites were worried about the arrival of Irish Catholics in the north-eastern cities.

But no matter how you cut it, the ethnic makeup of the American population is changing fast, and it will continue to do so. The Europeans arrived and decimated the indigenous peoples, and for centuries their demographic dominance was uncontested. But the nature of 'white' America subsequently changed, as the arrival of Italians, Poles and Jews challenged the demographic – and then the cultural – dominance of what came to be known as WASPs (White Anglo-Saxon Protestants). Many of the cultural and literary icons of mid- and late twentieth-century America, from Philip Roth to Madonna, were the descendants of arrivals from Europe's further shores.

However, since the immigration reforms of the 1960s, which reversed the restrictions of the 1920s, the United States has come to reflect a melting pot not of European ethnic groups but of the wider world. It is home to growing communities from Asia and Africa, as well as Latin America. Tomorrow's Americans are going to be very different from yesterday's – culturally, ethnically and religiously.

For many people, moving to the United States has provided great opportunities. But the journey is not easy, especially for those migrants who continue to risk their lives to come illegally. In a single month in 2019, 144,000 people were taken into custody while trying to cross the border. Hundreds are rescued from drowning in the Rio Grande, while many others are less fortunate.[12] In June 2019, the shocking photo of the dead bodies of a father and his tiny daughter became headline news. Óscar Ramírez had travelled from El Salvador with his family and was hoping to seek asylum in the US; he and his toddler had made it across the river, but when he went back for his wife, the infant followed and the two were swept away by the current.[13] Such tragedies are not new; around 1,600 migrants are believed to have died while trying to cross the Mexican border between 1993 and 1997 alone.[14]

What California has experienced is also being seen across the Western world, evidence that the expansive stage of the demographic transition has become a global phenomenon. Some white Americans once believed that Mexicans would 'melt away' before them, while British adventurers in Africa were concerned that the indigenous population might disappear when confronted with rampant European population growth. Charles Darwin, as we have seen, thought the 'civilized' races (meaning Europeans) would eventually wholly displace the others.[15] We can see the hubris of such sentiments in the Californian school data of today.

Demographic changes create the basic conditions for such change, as populations swell in the Global South and decline in the developed world. The burgeoning populations of the south are then attracted to the booming economies of the north. People tend to stay put when they are utterly destitute – only once they have gained some prosperity can they think about moving. And now, for the cost of a mobile phone, everyone can be beckoned by the seductive material prosperity of the developed world. This combination of demography and economics, which draws in immigrants and transforms the population, is happening all across the US and indeed back in Europe, to which we now turn.

### *Europe Transformed*

In the summer of 2015, a photograph of a small boy's body, washed up on a beach in Turkey, sent shockwaves across Europe. Just as four years later with the death of Óscar Ramírez and his daughter in the Rio Grande, a single tragedy came to represent a bigger human catastrophe.

Alan Kurdi was from Kobani, a town in northern Syria ravaged by fighting between Islamic and Kurdish forces. Having fled for the safety of Turkey, his family attempted to move on to Greece, but they barely made it beyond the Turkish coast. Alan's boat was one of two that capsized off the Bodrum peninsula on 2 September, drowning twelve people, including five women and children.[16] Like Óscar Ramírez and his daughter, three-year-old Alan became representative of the thousands who try to escape to the developed world and end up dying in the process. Europeans, like Americans, are faced with the agonizing challenge of controlling immigration and ethnic change on the one hand while also providing a safe refuge when it is needed.

An open-door policy, even if it were politically feasible, would encourage even greater numbers of ambitious or desperate people to make dangerous journeys, and would inevitably lead to more deaths. NGOs sometimes force European authorities to come to the aid of drowning women and children, but cries for help are often ignored. In August 2021, dozens of would-be migrants drowned off the coast of Africa while heading for the Canary Islands, a story so common that it only made the 'news in brief'.[17] Even when stories like these are widely publicized, immigrants keep coming – their miserable prospects at home contrast so greatly with the perceived opportunities in Europe that they are prepared to ignore any fear of the journey's dangers.

In 2015 more than 1.3 million migrants sought asylum in Europe, more than twice the previous year's number.[18] Many, like Alan Kurdi, came from Syria, fleeing a civil war that had been going on for nearly five years. Others came from places like Afghanistan, escaping longer-running conflicts and seeking economic opportunity in Europe. Since 2015, the tide has ebbed, partly as a result of a tightening of border controls. But to the vast young populations to its south and south-east, Europe's prosperity continues to act like a magnet. The recent takeover of Afghanistan by the Taliban is likely to trigger another such wave. Mass arrivals on the coast of Britain were once again hitting the headlines in the summer and autumn of 2021, when migrants and asylum seekers risked and often lost their lives in an attempt to get from one safe country to another, preferred one.

The ethnic makeup of Western Europe has been transformed by migration from two major sources: first, long-standing immigration from Africa, Asia and – in the case of the UK – the Caribbean; second, the more recent immigration from the former Communist bloc. This second inflow has come particularly from states that joined the EU and whose citizens are therefore entitled to free access. In 2018, around 6 per cent of residents in the UK had been born elsewhere in the EU, a significant share but smaller than the 9 per cent who were born outside Europe.[19]

The result of these two flows has been a noticeable change in the populations of Western Europe, and particularly in its big cities. The London where I was born, in the mid-1960s, was overwhelmingly peopled by those whose ancestors had resided in the British Isles for many generations. As the child of immigrant parents, I was unusual; a couple of decades earlier, before the arrival of the *Windrush* generation and early mass immigration from South Asia, I would have been even more exceptional.

By 2011, over a third of Londoners were foreign-born. In 2017, nearly 30 per cent of babies born in the UK had foreign-born mothers, a figure that rose to nearly 60 per cent in London. In the London Borough of Brent, where I was born, it was over three-quarters.[20] The figures for Paris, Brussels and Berlin are not dissimilar. Brent was overwhelmingly white in the 1960s, but by 2001, less than half the borough's inhabitants were white, and by 2011, barely a third of the population was.[21] I have little doubt that the 2021 census will show a further fall.

Such change can cause logistical problems, including for health services. In Charité, a hospital in Berlin, many women who give birth are from non-German backgrounds, which makes communication difficult. The director of obstetrics, Dr Wolfgang Henrich, is worried: 'Interpreter costs amounted to several hundred thousand euros this year, because of foreign women who simply showed up. I'm not only talking about Syrian women, but also about those from Iraq, Iran, Afghanistan or different African countries. In such cases, we need to provide interpreters on short notice. Financing these services is an unsolved problem.'[22] Schools and courts are facing similar challenges. The problems of multilingual societies, which in the past bedevilled the armies of Austria-Hungary and the Soviet Union, need to be resolved by modern welfare states.

Inter-continental migration was once a rarity – the distances and costs were too great, while the means of transport were too primitive. In the nineteenth century, when Europeans poured out of their continent, driven by a population boom at home and attracted by opportunities elsewhere, they were enabled by new forms of transport. Now, as Europe's demography contracts, this process of global migration has gone into sharp reverse. Rather than churning people out, Europe is sucking them in.

We have probably only witnessed the start of this process. Africa, as we have seen, is on course to burgeon demographically, so the pressure of people wanting to move to Europe will only intensify. Egypt, for example, is home to over 100 million people, more than live in Germany, having had less than a third of Germany's population in 1950. It is highly reliant on foreign financial support and aid; if this was disrupted, the wave of people seeking entry into Europe could dwarf anything we have seen so far. Much will depend on countries' immigration policy, but one expert has estimated that people of white British origin, who in the early 1990s made up over 90 per cent of the UK's population, will account for around 60 per cent by the middle of this century.[23]

On the other side of the English Channel, the picture is much the same. Although there is no official census data on religious affiliation in France, the surveys that do exist suggest that around 9 per cent of the French population is Muslim. Only a very small element is represented by indigenous converts;[24] the great majority are either immigrants from North Africa or their descendants. People from Morocco, Algeria and Tunisia who were familiar with French language and culture and lived in countries with high birth rates and underperforming economies were attracted to France, just as people from former parts of the British Empire were attracted to the UK.

As recently as sixty years ago, there was no substantial North African population in France, but there were over a million *pieds noirs*, people of European origin who lived in Algeria and were either evacuated when it gained independence or left shortly afterwards. The turnaround has followed decades of population growth among the indigenous Algerians, which radically changed the population balance between the two countries, sealed the fate of French Algeria and set the course for mass migration from North Africa into France. This would not have pleased the French statesman Charles de Gaulle, who suggested that it was only acceptable for France to have non-European minorities 'but on condition that they remain a small minority', lest his home town Colombey-les-deux-Eglises become 'Colombey-les-deux-Mosquées'.[25]

The survey of France's religious and demographic makeup cited above expects the country's Muslim proportion of the population to be 13 per cent by 2050. There will also be many Africans who are not Muslim, mirroring what will be seen elsewhere in Europe. Paris, London, Rotterdam, Frankfurt, Brussels and Marseilles are already home to significant populations who have either immigrated from outside Europe or who are descended from such immigrants. Tomorrow's Europeans, like tomorrow's Americans, will be very different people.

### *Migration versus Fertility: What is Driving Ethnic Shifts?*

Rapid ethnic change in a region can be driven either by differential fertility rates or mass migration, but differential mortality rates can also change the ethnic balance. This can be caused by genocide, and is also seen when immigrant populations are younger than locals. Whether or not they have a disproportionately high share of births, the age structure of immigrant groups makes them likely to have a disproportionately low share of deaths. When the proportion of Serbs in the Kosovan or Bosnian population fell after the middle of the twentieth century, for example, it had partly to do with their movement from those regions but was also due to the fact that Kosovars and Bosnians were having more children than their Serbian neighbours.

In the case of the United States, the major ethnic change since the early 1970s is more the result of migration than differential fertility rates. It is true that Mexico had a much higher fertility rate than the US for a period, which was a major cause of migration in the first place, but when immigrant groups move to places with low fertility, their rates tend to quickly converge with those of their new home. As we saw in Chapter 4, Latino fertility rates in the US have rapidly converged with those of whites. Indeed, the drop in Latino birth rates has been a key factor in the recent decline in US birth rates.

The straightforward reason for this convergence in fertility rates is the adoption of local ways of life by the young generation. And in the case of Mexico, fertility patterns are also shifting back home. If the birth rate is plummeting in Mexico and other Central American countries, it would be surprising if those who ventured north to live in America were not subject to the same modernizing forces.

Rather than urging the younger generation to procreate, mothers and grandmothers seem to be advising them to do the opposite.

According to Yoselin Wences, a daughter of Mexican immigrants to the US, 'The mindset was, "Don't be like us, don't get married early, don't have children early. Don't be one of the teen moms. We made these sacrifices so that you can get educated and start a career."' A student in South Carolina, she told an interviewer from the *New York Times* that she does not plan to have children until her mid-thirties.[26] Given the overall decline in Latino fertility rates, other young women seem to be receiving similar advice – and acting on it.

My own daughters went to a girls' school in London where most of the students were either immigrants or the children of immigrants, predominantly from South Asia. Their aspirations were university and a career rather than early marriage and a large brood of children. In this, they were not only conforming to British societal norms, but to trends we have already observed in South Asia. If ethnic change in North America and Western Europe continues, little of it will be accounted for by differential fertility behaviour of immigrants and their children.

The data supports this anecdotal evidence. It is not surprising that fertility rates for Indians in the UK have since the late 1980s been *below* that of the white British population. And people of Bangladeshi and Pakistani origin once had a higher fertility rate, but it markedly converged in the 1990s.[27] Many of these communities are relatively young, so they have more births and fewer deaths than the national average; even without further immigration, 'demographic momentum' means their numbers will continue growing for some time. But without further immigration, this effect will only deliver modest further change in the ethnic makeup of the UK.

We have already noted that fertility rates in remote rural areas of the United States are higher than urban areas. There was a time when some immigrants to America, from rural Scandinavia, for example, were attracted to the land, but immigrants to developed countries have long since overwhelmingly chosen to settle in towns. And having arrived in the city, whether they had come from cities or the countryside, they invariably start behaving like city folk. It might be envisaged that places in Europe and North America where the indigenous majorities live disproportionately in rural areas might have a higher fertility rate than is found among communities of urban immigrants; we can already see this in the US, where the fertility rate of white Americans is no longer markedly lower than that of immigrant minorities. Mormons in rural Utah, for instance, have many more children than Latinos in New York City. However, the differences in fertility between rural and urban areas in the developed world are generally slight and unlikely to result in any material ethnic shift.

### *Reversal, Reaction and Redefinition*

If the history of demography teaches us anything, it is that nothing is entirely inevitable. Before they took place, demographic events often seemed highly improbable, but having happened, they come to be viewed as having been inevitable. This is also true with regard to ethnic change – the future of North America and Europe is far from settled. Various forces will determine the outcome, and the choices made by individuals and politicians will be instrumental. In demography, free will rather than determinism reigns, even if it is the free will of millions of people that counts.

Economics and differences in birth rates and age structures may currently be encouraging more immigration, but it's likely that these trends will weaken as economics and demography converge. For example, Eastern Europe is getting both richer and older. Between 2005 and 2045, the number of Poles in their early twenties, a peak age for migration, will almost halve, reducing the pool of potential

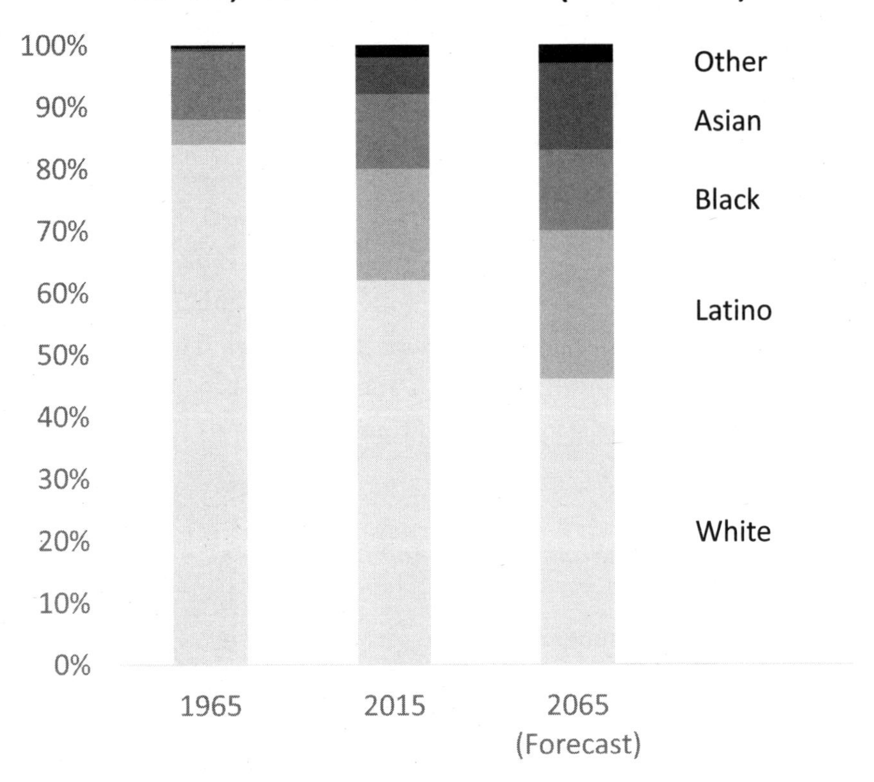


Source: Pew Center. Note, 'Asian' and 'Other' each below 1% in 1965

In the past, people of European extraction travelled to the furthest continents and changed their demography. Now this is happening in reverse, as the rich countries of Europe and North America draw in immigrants from Africa, Asia and Latin America.

Following mass immigration from Europe before the First World War and the subsequent introduction of strict controls, the US was very white when it changed its immigration policy in the 1960s. Since then, it has seen a particular influx from Latin America. By 2065, the white share of the population will have almost halved in a century, and whites will represent a minority.

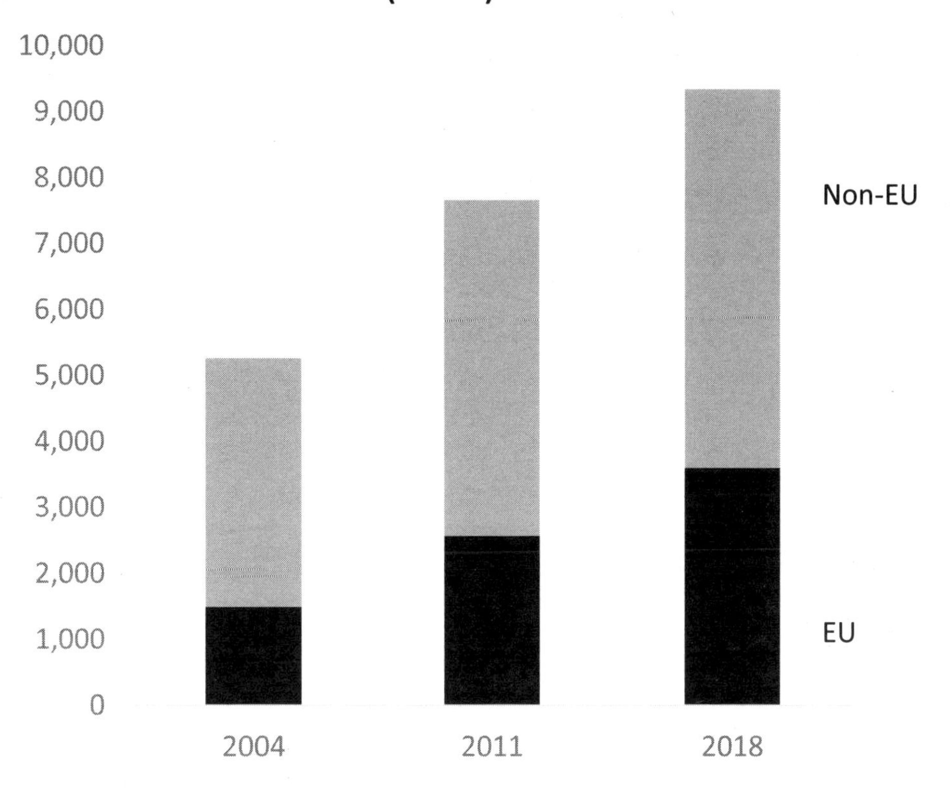


Source: Migration Observatory

With its insatiable demand for labour and its long-term low fertility rate, the UK has been a magnet for immigrants, both from within and outside the EU. The number of foreign-born people living in the UK almost doubled between 2004 and 2018.

migrants. The days of the ubiquitous Polish plumber in the UK may be coming to an end; with fewer young people entering the labour market, there will not be enough Polish workers to go around.

In the early 1970s, the time when Hispanic immigration to the US rocketed, Mexican women had nearly seven children each, while women in the US had just over two. Today, as US fertility has sunk even further, Mexican rates have nose-dived to a comparable level. This rapidly narrowing difference explains why the 'Mexican wave' in America is coming to an end, even if migratory pressure from elsewhere in Latin America persists. Meanwhile, as we have already seen, fertility rates among migrant populations usually fall, converging with those of the populations in which they are resident. And to some extent, since immigrants predominantly live in urban areas, their fertility rates may be lower than those of the nation as a whole.

In early twentieth-century America, the supply of Europeans must have seemed inexhaustible; it turned out to be anything but. Similarly, if the countries from where migrants are currently moving to North America and Western Europe in huge numbers experience shrinking demography and booming economies, this pattern will repeat itself.

Whatever the longer-term forces, short-term factors are already reducing immigration levels. Whether they are indigenous locals or relatively new arrivals themselves, people are frequently reluctant to accept immigration and ethnic change. Some 44 per cent of the UK population are in favour of lower immigration, according to a 2019 survey.[28] A few years earlier, at around the time of the 2015 European refugee crisis, a survey found that more than three quarters of British people held negative attitudes to immigration.[29]

Such attitudes can exist for years with little impact on a pro-immigration political consensus, but by 2015, even the Labour Party was calling for the introduction of more limits to immigration and attacking the Conservative government for failing to implement tougher controls.[30] The vote to leave the European Union the following year was, to a considerable extent, a consequence of negative attitudes to migration.[31] For many years, the UK government had committed to bring annual net immigration below 100,000, although it never achieved this target.[32] In 2018, over a quarter of a million more people entered the country than left it, with total arrivals at over 600,000[33] – far more people than had migrated to the UK than in the nine centuries between the Norman Conquest and the Second World War. The figures for 2019 were similar.[34]

On the other side of the Atlantic, Donald Trump's promise to build a wall on the Mexican border was his most memorable pledge in his successful 2016 presidential campaign. The typical Trump voter seemed more concerned about immigration and the changing American demographic landscape than about economic inequality or the failure of the financial system. But again, as in the UK, where there is a strong reserve of anti-immigration sentiment, policies to curb it are not only the preserve of right-wing politicians. Bill Clinton was president when the 1996 Illegal Immigration Reform and Immigrant Responsibility Act made the deportation of illegal immigrants from the United States relatively commonplace.

It's the same beyond the Anglosphere. The slogan of the Front National at the 2017 French presidential election was '*on est chez nous*' ('this is our home') – designed to speak to sections of the French population who felt overwhelmed by the presence of people whose cultures they considered alien. As the non-indigenous population of a country rises, so does electoral support for the far right. In 2017, Marine Le Pen's share of the vote in the second round of the French presidential elections was twice the size of her father's share fifteen years earlier.[35] And in the 2017 election, the leading candidate of the far left was also vociferously opposed to the free movement of people.

Italy's populist government in 2018–19 owed its electoral success as much to the fear of immigration from across the Mediterranean as it did to the country's economic woes. And the Austrian far right, its growth also spurred by migration fears, has recently participated in government. In Germany, the rise of the right-wing populist AfD pressured Angela Merkel's centre-right government into a much less accommodating stance on immigration; there has been no repeat of the 2015 influx.

However, we should be careful not to associate the rise of these anti-immigrant populist parties too closely with the fascism of Europe in the 1920s and 1930s. The key difference, once again, is reflected in demography. The new far-right parties are making gains in societies whose median ages are in the forties, not the twenties. Although they are conservative and resistant to rapid ethnic change, rightists in Italy and Austria are not forming street gangs. Indeed, one of the striking things about these populist movements is the absence of violence. If European democracy does wither in the face of the rightist populism, it will not be violently snuffed out as it was in the inter-war years. Europe's people are likely too old to take to the streets or to back movements that would be interested in launching military adventures beyond their borders.

Immigration and consequent ethnic change are, far from an inevitability, a choice made by government, which ultimately responds to public opinion. And although a state may find controlling its borders a challenge, it's not an insuperable one. Take Singapore, a prosperous island surrounded by Indonesia and Malaysia, which are much less wealthy. Although both these countries are making economic progress, they contain hundreds of millions of people who would be significantly better off if they could emigrate to Singapore. With a population of less than six million people, Singapore could be completely overwhelmed, but it patrols its borders with determination. Australia is equally firm with those who attempt to enter the country by sea, confining those who are caught to camps in the South Pacific. Countries in south-east Europe have erected fences to stem the flow of refugees from Turkey.

The progress that is being made in the developing world, as its fertility falls and its economy rises, is one way in which ethnic change might be arrested. At first, economic development makes migration to prosperous countries a target to which people who live elsewhere can aspire, but the creation of opportunities at home may persuade many to stay put. Wars are one reason people leave their country, and these are becoming less frequent. Additionally, minorities tend to be urban, so also tend towards low fertility; rural 'natives' have at least the potential for a higher birth rate, but the difference is likely to be slight, and the rural population is generally a small share of the total population.

A further factor that could limit – or even reverse – ethnic change is more subtle. Shifts in identities can be regarded as a sort of post-modern fuzziness of the boundaries between people that are thought of as absolutes, but are in fact more arbitrary than is often believed. This might sound a bit cryptic, so let me offer an example. When long-running ethnic tensions in Sri Lanka blew up into full-scale civil war in 1983, the world began to take note of a conflict between a Sinhalese majority and a Tamil minority. However, anyone who looks more closely will quickly realize that the situation was much more complex. Sinhalese people were an amalgam of high-country 'Kandyans' and low-country coastal people, who had different traditions. Until fairly recently they were counted separately in censuses. Tamils, meanwhile, included both 'Sri Lankan Tamils', long established in the north of the island, and 'Indian Tamils', the descendants of tea-picking immigrants from colonial days. Many Sinhalese and Tamils, rather than being Buddhist or Hindu, are

Christian. There are also Muslims, who are predominantly Tamil in language but not in affiliation.[36]

These complex, shifting and malleable Sri Lankan identities are built on myths that are often historically inaccurate. For example, it is widely believed that the Sinhalese are of north Indian extraction, but a leading Sinhalese ethnographer has suggested that a tiny Sinhalese core attracted a long-term inflow of south Indians who adopted their language and religion; most of those people who consider themselves Sinhalese are, in fact, genetically indistinguishable from the Tamils. 'Biologically,' he said, 'we are all Tamils.'[37]

There is nothing unusual about Sri Lanka in this regard – identities are generally less straightforward than people believe. Think of Ireland, for example. Many people there have a sense of having been oppressed by the English. But the people who moved to Ireland from England were in some cases Norman rather than English, and they often merged with the locals in different waves, so the Irish population today is more likely to be descended from such people than those in England. In Ulster, seventeenth-century Presbyterian settlers in the west of Ireland, often from Scotland, tended to assimilate and convert to Catholicism, while indigenous Catholics in the east, where incomers were more densely settled, often became Protestants. This explains how some Nationalist leaders have names like Adams and Wilson, while there have been Loyalist terrorists with names like Murphy.

Conversely, at the time when the IRA, many of whom were undoubtedly descended from English and Scottish immigrants to Ireland, were blowing up people on the British mainland, the UK had a prime minister named Callaghan and a chancellor of the exchequer called Healey, neither of whom identified as Irish. Callaghan's successor, Margaret Thatcher, who insisted that her instincts were 'profoundly Unionist', also believed herself to be of partly Irish extraction,[38] while Tony Blair, whose ancestors included Northern Irish Protestants, converted to Catholicism. Matters of identity, we can see, are just as complex among the peoples of the British Isles as they are in Sri Lanka.

On the other side of the Atlantic, the same phenomenon holds. We have already observed that one of the major issues that mobilized the Republican vote in the American presidential election in 2016 was Donald Trump's pledge to 'build a wall', but resentment against Latinos did not prevent his rivals for the Republican nomination having the names Cruz and Rubio. Latino identities fade with time and intermarriage[39] – today, for every two American Latinos who identify as Catholic, one identifies as Protestant and one is religiously unaffiliated, and the share of non-Catholics is rising.[40] In addition, regardless of religion, Hispanic identity wanes and intermarriage grows more common with each generation that follows immigration.[41]

Although the ethnic future of North America and Europe will undoubtedly be less European than its past, it is likely that many people who have arrived from further afield will come to integrate fully into the societies of their new homelands. That nature of those identities will doubtless shift over time, as they always have done; the thirteenth-century English were different from the tenth-century Anglo-Saxons. As mixed marriages lead to more and more people of mixed origin, it is quite likely that many people with little or no British ancestry will come to identify with the UK or one of its constituent nations, and the same thing will be true in other Western countries. The United States has been a tremendously powerful machine for creating more Americans; European countries could well be just as successful, although unlike the USA, they don't currently think of themselves as 'immigrant countries'. Much will depend on the rate of migration and the speed of integration.

From the perspective of London, Paris or New York in the early

2020s, the rise of an ethnically mixed society seems like a natural process, but this idea is historically inaccurate. At a point in history when the UK and France were ethnically homogeneous, cities in the Middle East, such as Algiers, Baghdad and Alexandria, were a mosaic of religions and nationalities. Today, in stark contrast, they are characterized by grim-faced uniformity or even ethnic segregation. There is no one-way path leading to a multi-ethnic future; to see one is an optical illusion that reflects a historically and geographically narrow frame of reference.

# 9

# Education

## *71: Literacy Rate Per Hundred Among Bangladeshi Women*[1]

This book has so far concerned itself with matters of *quantity*: how many people there are, how many children they have, their average age and at what age they die. All of this matters, of course, but it is now time to turn to questions of *quality*. For humankind has been undergoing the most extraordinary transition in its existence, a qualitative upgrade that knocks any technological advance into the shade. Without it, the great numeric changes that we have seen – falling death rates, rising life expectancies and falling fertility rates – would not have been possible.[2]

Put simply, humans have gone from being illiterate to being educated, with education going from the preserve of a vanishingly small minority to the entitlement of billions. From the perspective of tens of thousands of years of human history, all this has happened in the blink of an eye. In 1800, the world is estimated to have been nearly 90 per cent illiterate. Today we are fast approaching 90 per cent literacy.[3]

### *The Bangladeshi Educational Miracle*

On the eve of Indian independence in 1947, the country's Muslim leadership insisted on the creation of a separate Muslim state. They received a truncated territory that included what is now Pakistan in the west and Bangladesh in the east, the two wings separated by nearly a thousand miles. For a couple of decades, these two disjointed territories existed as a single country, despite their geographical and cultural differences. But when the Bengalis in the east could no longer tolerate the dominance of the Punjabis and other west Pakistanis, they revolted, and the west Pakistanis responded with what has generally been accepted as genocide. As many as three million Bengalis were killed,[4] with the Hindu minority a particular target of the Pakistani troops. Neighbouring India faced a huge influx of refugees. Its prime minister Indira Gandhi intervened on behalf of the rebels, and in March 1971 East Pakistan was separated from West Pakistan and became the independent state of Bangladesh.

The new state was born in inauspicious circumstances. The flat, rich alluvial Ganges delta provides excellent agricultural land, but the country's population had expanded by about 80 per cent in the previous twenty-five years, reaching the limit of its productive capabilities. The country was extremely poor and living conditions for the vast majority were basic. Just as Ethiopia was the poster child for suffering in the 1980s, Bangladesh filled this role in the 1970s, when it was struck by a series of cyclones and floods. Most of the country is very low-lying, making it prone to such events, while population growth had pushed more and more people into marginal areas, leaving them vulnerable when disaster struck. A major recipient of international disaster relief, Bangladesh was regarded as a classic 'basket case'; indeed, it has been claimed that the US National Security Advisor Henry Kissinger used this dismissive term.[5]

Although Bangladesh is still a poor country where life is hard for most people, the demographic data clearly shows that it is on a path to a very different future. Since gaining independence in the early 1970s, life expectancy there has extended from the mid-forties to the early seventies, while infant mortality has plunged to around a sixth of its early 1970s level. The fertility rate has fallen from nearly seven births per woman to barely two, bringing population stability into view, and all this has been underpinned by a revolution in literacy. As we have already seen, when you give people at least a basic education, they become better able to look after themselves and to care for their children, they live longer lives and they start to have smaller families. Bangladeshis are taking their fate into their own hands, and they are doing so by educating themselves.

The increase in literacy in Bangladesh marks the foundations of the process. Almost three-quarters of people there can read and write, with rates higher for men than women. But among those aged under twenty-four, literacy levels are well over 90 per cent, and higher for women than men. Literacy will soon be universal in Bangladesh, just as it is in countries like Canada and Japan. Education may not be a panacea, but social progress can only go so far without it – and mass literacy is the first step.

As with all data, the changing Bangladeshi literacy rates need to be put into a historical context. When Bangladesh became an independent nation, its literacy rate was not much more than one in four.[6] Among women it was barely one in six, which was less than half what it was among men. But by the second decade of the twenty-first century, the country's literacy gender gap had almost closed and literacy was approaching near-universal levels among the young.[7]

Literacy can be achieved with primary education, but the next challenge is to get more Bangladeshi boys and girls into secondary schools and keep them there at a time when economic pressures and social forces are pulling in another direction. Despite this, progress in the region at secondary level and beyond has been striking. In South Asia as a whole in the twenty years to 2014, the rate of enrolment in tertiary education quadrupled from one in twenty to one in five.[8]

The experiences of Bangladeshi girls who stay at school are transformative. 'My education . . . will make me a complete person,' insists Salma, an aspiring lawyer. 'My education will help advance this society.' Anjana expresses similar sentiments about self-development: 'If I think about not going to school, then I am very sad because then I cannot have a good life. It's important to learn and develop myself.' Like Salma, Rupa, who hopes to study medicine, wants to contribute to society: 'I want to be a doctor and stand beside the poor and those who are not receiving help,' she says.[9]

These sentiments represent the greatest force we know that will help humankind to escape the cycle of poverty and ignorance. It is hard to imagine illiterate women confined to villages and the drudge of child-rearing and agricultural labour having such lofty ambitions. As an adviser to the Bangladeshi government has put it: 'Women are better educated, safer, and more economically prosperous than their mothers. Today, women are accepted and valued not only as wives and help-mates, but also as farmers, parliamentarians, and entrepreneurs. The whole nation benefits.'[10]

### *Trailblazers: How Education Transformed East Asia*

In prioritizing education, the Bangladeshi authorities were following in the footsteps of East Asia. By the time of Bangladesh's independence, South Korea, Taiwan and Singapore were on a path of ultra-rapid development, which had education at its heart.

In the late 1940s, South Korea was one of the poorest countries in the world – and that was before it was devastated by war in the early 1950s. However, it managed twelve years of double-digit economic growth during the two decades to 1988,[11] and by the end of the twentieth century it was one of the world's most dynamic and successful economies. All this was achieved not through the exploitation of natural resources but through the intensive education of its population. Initiatives that were introduced included universal access to free school education and increased pay for teachers, nearly a third of whom have master's degrees. As a result, enrolment in college and university rose from 30 per cent in the mid-1980s to over 95 per cent today.[12] According to rankings conducted by the OECD's Programme for International Student Assessment, South Korea is among the world's top ten countries for reading, science and maths, categories in which neither the US nor the UK make the top ten.[13] And South Korea's success is evident in the economic and demographic data, whether in its status as the world's tenth largest economy or its astonishing decline in infant mortality and its significant extension of life expectancy.[14]

Some people question whether education is really the engine of economic transformation, suggesting that it might be an effect of prosperity rather than a cause. The answer, surely, is that development and education go hand in hand. It is hard to envisage a prosperous and modern society like South Korea, full of highly productive people doing complex jobs, not having an educated population. Perhaps nineteenth-century Britain was able to advance towards industrialization without most of its population being educated, but that is hardly likely now, at a time when well-paid jobs are so much more intellectually demanding. And it is notable that living standards in Britain only indisputably started to rise in the 1880s, the decade *after* the introduction of compulsory primary education.

The data shows conclusively that there is a strong correlation between educational attainment and economic achievement, when comparing both individuals and countries. One US study, for example, showed that people who have an advanced degree had more than five times the income and more than eighteen times the wealth of people without a high-school diploma.[15] If education simply positioned people to take a job at the top, rather than expanding the number of such jobs, it could not explain how whole societies have managed to go from poverty to prosperity, as we have seen with South Korea.

Education might be a necessary condition for economic progress, but it is not the only one. After all, it is possible to have an educated population and yet fail to make the economic grade. In many countries in North Africa and the Middle East, people graduate from university, only to find themselves unable to get a job. The quality of these degrees is often poor. For example, Egypt, by far the most populous Arab country, has a university sector that is ranked by quality at 130 out of 137 countries.[16] Many of those with sufficient skills try to leave.

Where a country fails to integrate itself into the global economy and cannot offer well-paid jobs to well-educated people, investment in education tends to wither, as neither individuals nor the state is incentivized to make further progress. In Egypt, for example, unemployment is higher among graduates than non-graduates, as the labour market is not able to meet the expectations of the graduates emerging from substandard universities.[17] The creation of worthless degrees and unrealistic expectations is one of the greatest forces for instability, especially in countries where the youth constitute a large proportion of the population. The frustration of graduates, nearly half of whom are unemployed,[18] was a major cause of Egypt's 2011 uprising. The situation in nearby Lebanon is similar: thirty-five thousand graduates emerge each year, but there are jobs for only five thousand of them.[19] An excess of educated people relative to their economy's needs and opportunities has been blamed for revolutionary movements in the West since at least 1848.[20] And although Bangladesh's progress in terms of the lower levels of education has clearly aided its rapid economic growth, it also struggles to employ its graduates.[21]

## *Women, Education and Development*

While gender equality still has a long way to go, nowhere is the progress that has been made more striking than in education. In many countries, women now outnumber men in universities, sometimes by a wide margin. Seven Icelandic women go to university for every four men, while Kuwait has more than twice as many women as men in higher education.[22] In countries like Iceland, this is part of broader female emancipation, which can be seen in the workplace and in politics. In countries like Kuwait, however, it will likely give rise to a build-up of frustration, as increasingly well-educated women find themselves with limited opportunities in business and public life. On current trajectories, however, female equality will inevitably pass from education into public life. Bangladesh has been ruled by a female prime minister for more than half of its existence.

But however high women climb in business and politics, the simple fact of increasing female literacy has already proved transformative in poorer countries. Female education is one of the most effective ways of lowering fertility rates. Literate women are more likely to be able to take care of their own bodies and those of their children, resulting in lower mortality, particularly among infants. It is also likely that they will ensure that their children are educated at least as well as they were, which creates a virtuous inter-generational circle. The aspiration to improve one's position through education is one of the most powerful forces that encourages countries to progress through the demographic transition, driving down fertility and mortality rates. Education is also, of course, an end in itself, being the pathway to greater agency and a more satisfying life.

It has become an orthodoxy in international development that focusing efforts on women is an effective way of pulling countries out of poverty. As Larry Summers, the former chief economist at the World Bank, said, 'Investments in girls' education may well be the highest return investment available in the developing world.' Not only do such investments yield direct economic returns, they also pay dividends in the home, both in terms of the welfare of the next generation and also in moderating their number. Furthermore, women are still educationally disadvantaged in many places. Because they are often the most marginalized in poor societies but also do most of the work, and because they are responsible for the next generation, investment in their education pays the biggest dividends.[23] In Africa, women are estimated to own 30 per cent of the land but to produce 70 per cent of the food.[24]

Aside from being good in itself, education yields economic returns. Educated people are more productive, more able to progress to occupations that create higher value and more likely to participate in the formal economy. Education allows a poor subsistence farmer to adopt new farming techniques that improve yields or to take a job in a factory. It both provides access to information and allows that information to be used more effectively, while also improving a population's ability to adopt new technologies. There seems to be a global correlation between education and agricultural productivity; one study suggests that an additional year of education will result in more than a 3 per cent increase in output.[25]

Education not only makes farmers more productive; it gives peasants the basic level of literacy required to handle machinery or to work in a factory. And this has been the main driver of China's economic emergence since the 1980s.

The economies of all developed countries are reliant on women occupying positions that span a breadth of capabilities and competences. Our hospitals, boardrooms and parliaments would be unworkable without women – yet none of this would have been possible without the expansion of educational opportunities.

## *Education and Democracy*

If education's economic benefit is productivity and development, its political benefit is generally accepted to be democracy. However, this claim can be contested. India has existed as a democracy for decades during which the majority of the population was illiterate. Things have changed now, of course – literacy has risen from less than one in five to more than three in four since the country gained independence in 1947.[26] But even in the days of mass illiteracy, Indian democracy somehow survived.

On the other hand, there continue to be many countries that are, at best, only partially democratic, despite having well-educated populations. The Soviet bloc contained some of the best-educated people in the world. And China's population has made extraordinary educational advances without adopting a Western political system, showing that high levels of education can exist without democracy.

However, there is a long-standing view that whichever way the causality works, education and democracy are linked. When Britain expanded the right to vote in the nineteenth century, the upper classes, realizing that power was passing to the masses, were keen that they be able to exercise power responsibly. It is no coincidence that just three years after urban male workers had been given the vote in 1867, compulsory primary education was instituted – under the slogan 'we must educate our masters'.

Statistically robust studies have investigated the correlation between democracy and education and there does seem to be a link, although given how contestable the terms are, it is not surprising that there is no definitive proof.[27] This might seem like an obscure academic argument, but it is not. The question is whether modern, prosperous and educated societies need to take on a democratic character, or whether countries can exist with people who are equipped to participate in the global economy but accept their exclusion from political decision-making. No question is more important to the political future of humankind.

### *Education and its Discontents*

So far, so uncontroversial. Education is clearly a good thing, both in its expansion of individuals' horizons and as an instrument for economic development and demographic progress. The link between better education and longer life expectancy holds in developing countries where the literate are better able to care for themselves and their families. It also holds in the developed world, where mortality rates are distinctly higher for those without an undergraduate degree.[28] It even seems to create the conditions that are conducive to democracy.

This has to be preferable to a regime where life is prolific but cheap, and where most people do not participate in the political process. Education furthers development; in a sense, it *is* development. Along with per capita GDP and life expectancy, it is one of three measures used by the United Nations to calculate the Human Development Index, a measure of human wellbeing.

Some sceptics argue that causality flows from prosperity to education rather than the other way around – that it is the rich who can afford education rather than education that makes people rich. This may be wrong, but clearly not all education is good or represents value for money, whether it is paid for by the state or the private market.

Even where education *is* good, it doesn't always meet the requirements of the marketplace. We have already seen how it doesn't necessarily advance earning potential in the Middle East, and another example can be found in rural south-west China. Minority groups are obliged to attend school, but researchers have found that some parents withdraw their children from the classroom and make them sell vegetables instead, claiming that even college graduates struggle to find work in their remote mountain villages. At best, children with an education would end up with factory work, while one parent claimed that it might make them too lazy for farm work.[29] This particular father may not be making an accurate judgement, but education is not valuable, either at a personal or national level, if the circumstances aren't right and if the opportunities that should follow don't exist.

On the other hand, a hunger for education drives many people forward, despite the sacrifices that are required. Earlier generations of immigrants to America worked long hours to give their children opportunities they never enjoyed themselves. They saw education as a ladder; although they could not climb it, they ensured that future generations could. This is not a uniquely American or a uniquely immigrant phenomenon – all over the world, youngsters go to school to get the best education possible, often without shoes and on empty stomachs. 'I don't worry that I don't eat in the morning,' says Nyirenda, a schoolboy in Malawi, 'because I believe in the future when I will be a businessman, I will have more food.'[30]

Another attack on education asserts that it is designed to turn humans into production units for modern industrial capitalism. The science-fiction writer and futurologist Alvin Toffler wrote that 'mass education was the ingenious machine to produce the kind of adults it needed . . . the regimentation, lack of individualization'.[31] This criticism has followed education from industrial society into the post-industrial era. Now the complaint is that people are being de-skilled and trained to act as automata until technological advances render their jobs superfluous.[32]

The need to achieve and conform is a cause of psychological pressure that not all can bear, and South Korea is a case in point. Although education has powered the country to heights of prosperity that were previously barely imaginable, the results-focused system causes a great deal of anxiety in the country's young people, as they struggle to gain a place at an elite university by passing highly competitive entrance exams. No fewer than 86 per cent of pupils say they feel stressed, almost 75 per cent admit to feeling guilty if they take a break and the typical school student does not stop working until 11pm. 'I watch what my other friends are doing. Then I feel guilty, and so I feel I need to study some more,' says one.[33] South Korea has the highest suicide rate of all countries in the OECD, as well as the world's highest suicide rate among those aged between ten and nineteen.[34]

In many developing countries, Western-style education was introduced by colonial authorities, so it is condemned by some as a white European invention that displaces indigenous forms of knowledge. But many leaders of developing countries are more pragmatic and see education as the most reliable route *out of* dependence on the West.[35] It is up to each country how modern learning is combined with local culture and traditions. In this respect, Japan is an exemplar: it is a highly successful adopter of Western science and education that preserves its own ancient traditions and distinctiveness. And as more and more of the world becomes more educated, there is every reason to believe that knowledge will come to be seen less as a Western preserve and more as a joint human enterprise.

At its worst, it is asserted, education not only churns out mindless automata but breeds nationalism and even genocidal sentiment. We are often reminded that the Germans of the inter-war years were among the world's best-educated people and yet they supported the Nazis and ditched their democratic institutions in a headlong rush to war and mass murder. The 1994 Rwandan genocide is often blamed on the role of anti-Tutsi indoctrination in schools.[36] If nationalism is a problem, the role of education in propagating it has been central from the beginning,[37] and religion is propagated today as much through the school as the home.

However, these criticisms are akin to faulting motor cars for creating pollution, causing accidents and being used in bank-robbery getaways. The answer is that while education *can* be put to bad uses, it is far more usually put to good ones. Modern nation states and industrial economies could not have come into existence without the standardization of language and outlook, as well as the numeracy and literacy of their citizens. And just as the traditional nation state could not have developed without education, globalization also requires education in order that people see themselves not only as citizens of particular countries but also of the world. The most effective riposte to anti-education arguments is to ask whether we would be better off if everyone remained in ignorance.

A more interesting critique is made by David Goodhart and Dietrich Vollrath, who have separately made the point that the most advanced economies may have reached 'peak education'. In economic terms, when we are sending half the population to university, we have a low return on investment in human capital.[38] The graduate premium in the most developed countries has declined.[39] Goodhart asserts that we should think about what we value and reward in the workplace, while for Vollrath, it is an explanation for the slower economic growth of recent years.

Developed countries should contemplate both approaches, but humanity as a whole is still some way off facing this problem. As long as the majority of people – and even the majority of young adults – in a country like Chad are illiterate, over-education remains a First World problem.[40] And although fantastic progress has been made in countries like Bangladesh, there is much further for these countries to go, while others have barely started.

Indeed, while we should note the success of countries like South Korea and Bangladesh, it is concerning that many states are failing to progress educationally. We have already cited Chad, but it is far from alone in sub-Saharan Africa; although it is declining, the region has a greater gender gap in literacy than anywhere else in the world.[41] Still, the progress that has been made is clear from the fact that three-quarters of the young are literate, compared to a third of the region's elderly. But while the general picture is one of improvement, the vast challenges are being made all the greater by the population explosion taking place. In the small west African country of Equatorial Guinea, for example, while illiteracy is falling as a share of the population, the number of illiterate people is increasing because the population is growing so fast – it has almost tripled since the mid-1990s.[42] Providing universal education is difficult when the number of young people is rising exponentially.

## *The Great Knowledge Project*

The idea of education for all is so prevalent in much of the world that we take it for granted. However, it is sufficiently recent and historically unusual that its consequences will be huge.

When students at the universities of the West complain about the predominance of dead white males who appear in the curriculum, they have a point; if there was an eighteenth- or twentieth-century African woman equivalent to Isaac Newton or Albert Einstein, we ought to know about her. But the absence of such figures from the reading lists of university physics departments is not evidence of male or European superiority, as some on the far right claim, but neither does it stem from prejudice on the part of those who set the university curriculum.

Rather, the predominance of dead white men in our culture is a reflection of the people who had access to education and knowledge in the past, and of the people who did not. Newton and Einstein were exceptional individuals and had to struggle to develop their ideas, but at least they were able to do so. Until recently, the only people who enjoyed such status were white men. For the vast bulk of humanity over the vast bulk of history, even the most rudimentary education was not available. For most people, anything beyond basic literacy was a luxury they simply could not afford.

However, a great change is now underway. As the president of Armenia put it, 'If you can find Newton in one thousand and Einstein in ten thousand, imagine how many more talented people you can find in hundreds of millions.'[43] What's more, modern communications technology has vastly intensified the ways in which people can collaborate within and between different fields. International conferences have been facilitated by the ubiquity of air travel – or at least they were until the Covid-19 pandemic. And a variety of methods for sharing information, from email to Zoom video conferencing, means more powerful and dense networking. The result of many more educated people communicating ever more intensively is that knowledge accumulates much faster.

Two examples illustrate this point. After attending a concert with a friend of mine, an authority on nationalism, I suggested that he write a book about music and the nation. He protested that despite his musical enthusiasm, he lacked the necessary in-depth academic knowledge. I contacted a musicologist whose work on Elgar we had both enjoyed, and the result was a book jointly authored by my friend and the music professor, despite the fact that they only met in person once or twice. In any earlier age, they would have found exchanging ideas a slower and more arduous activity. It gave my friend purpose in the final years of his life, and it gave the world a fascinating study of a neglected topic.[44]

A more prominent example has been the development of treatments and vaccines for the coronavirus that has ravaged the world during my writing of this book. At the time of writing, the story is far from over, but the progress that scientists have made in decoding

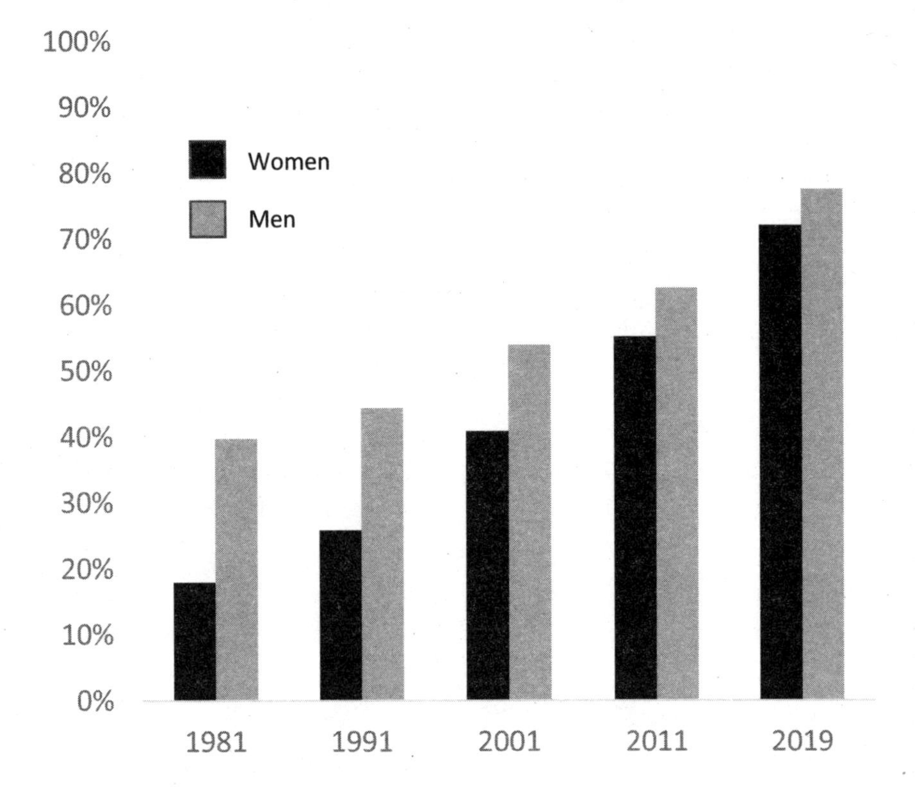


Source: World Bank

Since the early 1980s, basic education in Bangladesh has become near-universal. As a result, almost three-quarters of the country is now literate, compared to barely 30 per cent forty years ago. The other great success story has been the narrowing of the educational gender gap. In 1980, there were twice as many literate men as women, but the gap has been closing in the decades since then.

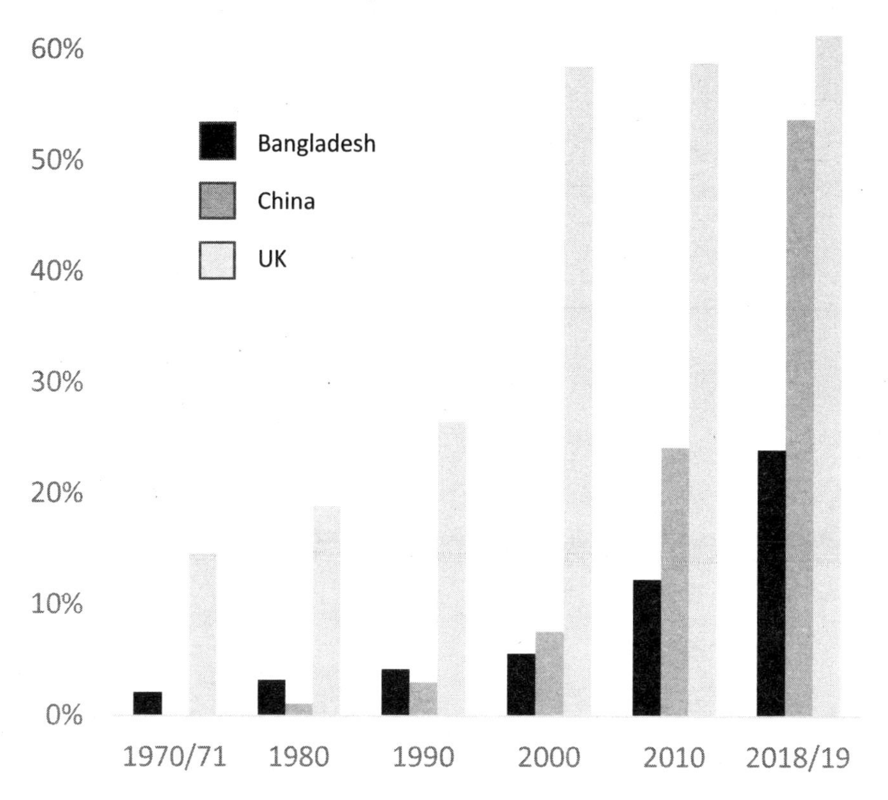


Source: World Bank

The rise of tertiary education is a global phenomenon, and an area in which China has made extraordinary strides. During Mao's Cultural Revolution, barely any Chinese people enrolled in college or university; now, about half do.

the virus and finding ways to combat its effects has been far quicker than would have been the case without both the sheer number of people working in the field and their ability to communicate and share results. A hundred years ago, a Cambridge scientist would have worked alone in his laboratory, perhaps writing the occasional letter to a colleague in Germany and waiting weeks for a response. Progress would have been at a snail's pace compared to what can be achieved today, when thousands of people across the world are able to interact with each other in seconds. Meanwhile, the Covid-19 pandemic has been less deadly and less disruptive as a result of the information and telecommunications technology that allows so many people to work and interact without having to travel or meet face to face.

### *Tomorrow's Educated People*

In the 1960s and 1970s, many academics worried that the world's population would eventually outstrip its resources. With global population growing at 2 per cent a year, such anxieties about the ability of the human race to feed and water itself were understandable. However, with population growth now at half that level and continuing to fall, and with some places suffering steep population decline, some people now worry that there will eventually be too few people. As we've seen, the size and age of the workforce has a significant economic impact, not least in its effect on the number of consumers, and if the workforce is shrinking, so might the economy. But there are balancing forces; while the number of new workers is likely to fall across the globe – as it has already in many places – their productivity is set to rise. China's recent economic miracle has been driven by the transfer of low-productivity agricultural workers from the farm to the factory, where their economic output rises. And as more people climb the educational ladder, there will be more potential for this kind of rising productivity.

Global economic growth will increasingly depend on raising the quality of the workforce in the face of its declining quantity. Future economic growth and development will have less to do with the number of low-skilled workers and more about the replacement of those who currently work with their hands by those who work with their minds. At the same time, artificial intelligence will offer an opportunity to replace much existing work. AI is most likely to replace many white-collar jobs rather than those that require manual dexterity or empathy; the bookkeeper has more to fear from the 'rise of the robots' than the dustbin man or the carer for the elderly, because emptying bins, changing bedding and providing empathy are extremely difficult skills for machines to replicate.

In more advanced countries, where an ageing population is already driving a huge demand for workers in the compassionate professions, the education of the population may have reached its high-water mark. There are still hundreds of millions of people across the world who are a long way from reaching their full potential, but that is changing fast.

# 10

# Food

## *375: Percentage Increase in Cereal Production in Ethiopia in the Last Twenty-Five Years*[1]

Imagine that all women since the start of the Common Era had reached their childbearing years, survived them and had an average of four children each. Imagine further that the average age of a woman giving birth was twenty-five. These appear to be fairly modest assumptions; a woman who is sexually active throughout her fertile years would typically expect more than four pregnancies, four live offspring does not seem that high a number and twenty-five as an average age for childbearing in a pre-modern society is not particularly young. However, these apparently conservative assumptions would make each cohort double the size of the previous one, and would have meant four doublings per century.[2]

If we apply these assumptions, starting from the year 1 CE, when there were around a quarter of a billion people in the world, the population would have grown to more than 250 thousand billion by the year 500 CE – more than *thirty thousand times* its current size. And by now, early in the twenty-first century, it would have had thirty-three digits rather than ten. These are the sorts of numbers that are usually dealt with by cosmologists or mathematicians, not demographers or social scientists. At some point, as one demographer suggested, the human population would have been expanding faster than the speed of light – eventually there would be more people than atoms in the universe.

The very idea of such population growth is absurd; a population expanding at the speed of light would hardly be in a position to procreate, and since we each contain many atoms, we cannot collectively outnumber them. But it is not immediately obvious at what point humans would reach the limit of what can be achieved. To have four surviving children is not such a big deal, as we have already seen, and the doubling of a cohort from one generation to the next is something that has happened quite often. However, the point is that it has not gone on for centuries.

Far from this kind of exponential growth, the increase in human population throughout history has been slowed by wars and pandemics. But the greatest constraint of all has been a lack of food – the planet could never have provided for anything like so many people. Long before running out of space, people would run out of food to eat. So as Thomas Malthus, the father of modern demography, explained, either starvation or war or calamity would hold back this great human surge, or sexual abstinence and infanticide would be needed to control it.

Modest though the above assumptions concerning our reproduction might have appeared, humans have consistently and massively undershot them. Populations expanded, only to be knocked back by disasters or suffering, including hunger, starvation and massacre. A huge number of births were required just to maintain a stable population. As one character in Pearl Buck's *The Good Earth*, a novel set in China, laments to his son: 'Ah me, to think that out of all the children I begot and your mother bore, one after the other – a score or so – I forget – only you have lived! You see why a woman must bear and bear.'[3]

But in the two centuries since Malthus, his two fundamental assumptions have been turned on their heads. As we have already seen, the human tendency to reproduce has been curbed. At the same time, our ability to produce food has grown not incrementally, as Malthus had expected, but exponentially. The production of food was the greatest demographic constraint, and the removal of that constraint has been one of the fundamental drivers of modern demographic change.

### *Ethiopia Escapes the Malthusian Trap*

In an Ethiopian hospital a hundred miles south of Addis Ababa, a smiling health worker weighs a robust-looking infant called Bontu and observes that he is developing into a well-nourished child, as the boy's mother looks on with pride. It's a scene that is now common in most of the developed world, but for a long time it was all too rare in sub-Saharan Africa.[4]

In the mid-1980s, Ethiopia was struck by famine. As is so often the case with such catastrophes, it was the result of a combination of natural causes (in this case drought), government incompetence (agricultural policies inspired by Soviet Marxism) and government malice (an attempt to damage rebel ethnic groups). As a result, around a million people died and life expectancy at birth fell to a staggering six years.[5] The West took notice of the disaster; those of us who recall that time remember the pictures of emaciated children too malnourished and exhausted to brush the flies from their faces. As a result, a generation of Europeans and North Americans came to see Ethiopia as the ultimate example of economic failure and human need.

Yet today, Ethiopia has been transformed – and babies like Bontu are the beneficiaries. Since 1984, while the country's population has more than doubled, the proportion of infants who don't live to see their first birthday has plummeted to less than 5 per cent, roughly half the level at the start of the twenty-first century and a quarter of the level at the time of the famine. The average daily calorie intake in Ethiopia rose between 1984 and 2011 from 1,500 to a much healthier 2,100.[6] Life expectancy has lengthened astonishingly, from forty-four to sixty-four since the early 1980s, while maternal mortality has fallen by two-thirds over the same period. Moreover, the adult literacy rate has doubled since the mid-1990s, from around a quarter to around half the population.[7]

The main way in which this has been achieved has been through the removal of the malign influence of Marxist-Leninist orthodoxy on agricultural organization. The support of the international community has been important, too; the facility where Bontu was weighed is funded by Canadian aid. Yet just as important have been the techniques and technologies Ethiopians have learned themselves, and the ways in which they have adapted them to local conditions. The most direct impact has been in agriculture and the growth in output mentioned above.

Although in Ethiopia life is still tremendously precarious for many people – made more difficult still by the recent re-flaring up of civil war – over the past three decades it has seen dramatic improvement in human wellbeing. None of this would have been possible without improved agricultural output. It is the opposite of what Malthus expected would happen – Ethiopia has significantly reduced its population growth rate while simultaneously achieving exponential growth in food production. In some places, yields more than doubled in just three years. Despite this, the country's overall wheat yields per hectare are still less than a third of those in America,[8] millions remain vulnerable to crop failure and some Ethiopians continue to suffer from malnutrition. Some 38 per cent of children had stunted growth in 2016, but the figure was 58 per cent just sixteen years earlier.[9] Despite Ethiopia's booming population, the share of undernourished people has fallen since the start of the century from more than 50 per cent to around 20 per cent.[10]

If food production grows four-fold in a generation but

population merely doubles, then each successive cohort will have twice as much food per head. While this level of increasing food production is clearly not sustainable over the long term, other countries have made gains on this kind of scale in recent decades. In the Indian state of Punjab, wheat and oilseed output grew by about 5 per cent per year for the forty-five years to 2005, a nine-fold increase.[11] On a global scale, grain production tripled in the second half of the twentieth century, and it rose again by almost 50 per cent in the first eighteen years of the current century.[12]

One of the constraints on increasing food production is, of course, the environment. It is often in poor countries with rapidly growing populations that the most environmental damage is done. Having once been heavily forested, by the early 1990s Ethiopia's tree cover was down to 3 per cent of its land area.[13] In 2019, the government claimed to have planted 350 million trees in a single day; there is some scepticism as to whether this was really achieved,[14] but it is clear that significant reforestation is taking place and that Ethiopia is starting to repair some of its environmental damage.

It will be difficult to improve Ethiopia's environment while its population continues to grow, but two factors will work in the country's favour. First, there has been a shift away from using wood as a fuel to alternative energy sources, including hydroelectric power from the Blue Nile. The second factor is a reduction in the pace of population growth. It peaked in the early 1990s at 3.7 per cent per year and is now barely above 2.5 per cent, and it should fall below 2 per cent some time in the 2030s. Although this will still mean more mouths to feed, the end of exponential population growth is at least in sight. By the end of the century, UN median estimates suggest that Ethiopia's population will have stabilized at around a quarter of a billion – still a great deal more than the current 100 million-plus. The country's fertility rate has already fallen to below 4.5, having been almost 7.5 in the early 1980s. And in

Addis Ababa, the fertility rate appeared to fall below replacement level as early as 1994.[15]

In many parts of the world, and particularly in the parched and increasingly populous Middle East, lack of water is a limitation on agriculture. But here, too, there are technical solutions. The cost of desalination of seawater has dropped dramatically in recent decades, and it already provides half of Saudi Arabia's drinking water.[16] Like any other fix, it creates its own environmental problems, but these are also being dealt with.[17]

### *Feeding the World: The Great Innovations*

Despite widespread concerns about the environment, the depletion of resources and climate change, we should consider how the planet can contain so many people and still be able to feed them all, in theory if not in practice.[18] Some may regret the existence of so many people, but whether or not population growth is welcome, population decline has already begun in some regions, and it is spreading. With the pressure starting to fall, it is time that we appreciated the innovations that have enabled the existence of human life on a scale that was unimaginable.

At the end of the nineteenth century, there were solid grounds to believe, as Thomas Malthus had at the century's start, that mass starvation lay ahead. It was true that modern agricultural production had expanded into the New World, and methods of transporting it had been developed. Yields had risen, and exports of beef and pork from the US grew fourteen-fold between the early 1850s and the late 1890s.[19] US wheat exports expanded rapidly from the 1840s, while the price of bread in Britain halved between 1840 and 1880.[20] The population of Britain had more than tripled since the publication of Malthus's *An Essay on the Principle of Population* almost a century earlier, and millions of Britons had also settled abroad. Improved agricultural techniques and food from beyond the continent percolated across Europe, so other countries' populations, and not just that of Britain, began to grow.[21]

As the nineteenth century drew to a close and the twentieth century dawned, it appeared that population in Europe had risen as far as the new frontier of production would allow. Food was being imported into Britain in particular from beyond Europe on a massive scale. Between 1850 and 1909 Britain went from being largely self-sufficient in wheat to importing 80 per cent of the wheat used in bread.[22] It seemed as if the gains in food production were over, with no more 'ghost acres' that could be put to productive use. Malthusian theory still applied, but from a position where there was more food and people. No new Americas could be discovered and there were no new Great Plains to be settled – those gains were already in the bag. Natural fertilizers could raise production only so far. In fact, deposits of saltpetre were so highly prized in Latin America that competition for their control led to vicious wars between Chile, Peru and Bolivia between 1879 and 1883, in which over 55,000 men were killed or wounded.[23] And whoever controlled and profited from the natural fertilizer, it was reckoned that only thirty or so years of supply remained.

It was against this background that William Crookes, president of the British Association for the Advancement of Science, set out his hopes that science would find a way around these constraints. Such a breakthrough would indeed be made across the North Sea in Germany.

In the years before the First World War, the German chemist Fritz Haber developed a process of nitrogen fixation that was subsequently scaled up by Carl Bosch. This allowed the creation of artificial fertilizer, thus ending the reliance on saltpetre and other natural deposits. As one of his eulogists put it on his death in 1934: 'Haber will go down in history as . . . the man who won bread from the air and achieved a triumph in the service of his nation and all of humanity.'[24] A more recent commentator noted that 'The world's population could not have grown from 1.6 billion in 1900 to today's six billion without the Haber–Bosch process,' which he called 'the detonator of the population explosion'.[25] According to current estimates, 40 per cent of the world's population is fed thanks to Haber and Bosch.[26] Although the population explosions in Asia and Africa in the last decades have depended on these two men, the vast majority of us do not even know their names.

Haber's approach was adopted by the Nazis, who needed to feed their population in the Second World War, even though the idea of solving food shortages using innovation rather than seizing land from others undermined the thrust of their programme.[27] Apart from Haber's Jewish origins and the implications of his achievements that made the Nazis suspicious of him, they also preferred a more organic approach to agriculture.[28] After their rise to power, Haber escaped first to Britain, where he was given sanctuary despite the contribution his invention of poisonous gas had made to the German First World War effort. He was travelling to British Mandate Palestine when he died of natural causes in 1934.

Haber's achievement was far from the only innovation that allowed agricultural production to increase, but it was essential to the world's ability to feed a population of seven billion people. Great advances were also made in combating weeds, insects and fungi, and these contributed to boosting yields. Another highly significant development was the so-called Green Revolution between the 1930s and 1960s that involved the adaptation of crops such as dwarf wheat and IR8 rice, which allowed some yields to double in the space of a couple of decades. The name most closely associated with the Green Revolution is that of the American agronomist Norman Borlaug, who developed new disease-resistant crop strains.

Like Haber, Borlaug won a Nobel Prize, but in his case it was in recognition of his contribution to peace – his innovation proved that human collaboration and creativity could overcome bottlenecks in productivity, and that history need not be an endless struggle between individuals, races or classes. The descendant of Norwegian immigrants to the US, Borlaug undertook most of his work in Mexico, but it had its greatest impact in India. For the Americans, a scientific answer to world hunger, and particularly one that had been developed in America or at least by an American citizen, played a useful political role, easing mass hunger and poverty among the potentially angry and revolutionary Third World peasantry.

There are those who say that Borlaug's innovations reduced genetic diversity and led to soil erosion, and he himself was aware of the limitations of his work.[29] But even his critics cannot deny that his ideas, like those of Haber before him, were instrumental in saving billons of lives.[30] As Borlaug noted, his critics had probably never 'experienced the physical sensation of hunger. They do their lobbying from comfortable office suites in Washington or Brussels. If they lived just one month amid the misery of the developing world, as I have for fifty years, they'd be crying out for tractors and fertiliser and irrigation canals and be outraged that fashionable elitists in wealthy nations were trying to deny them these things.'[31]

## *Different This Time?*

Food production has expanded vastly in the past two hundred years. Famines were still far from unusual in Europe in the nineteenth century, and they remained common in much of the world well into the twentieth century in, for example, Yemen, Sudan and Somalia. In India, devastating famines occurred as recently as the 1940s, with the 1943 Bengal famine resulting in the death of more than three million people.[32]

Increasingly, starvation is the result not of a lack of food but of war, political incompetence or design. There was no agricultural reason for millions of people to die of hunger in the Ukraine in the early 1930s; either the government intended to decimate the peasantry or famine was the inevitable outcome of the dogmatic design and implementation of collectivization.[33] The Ethiopian famine of the 1980s was, to a large extent, the result of an emulation of the Soviet model and of ethnic conflict,[34] which brings to mind a joke of the time based on the name of a now-defunct Communist magazine: 'Marxism Today. Famine Tomorrow.'

Death by famine has become much less common since the 1960s. One estimate is that the annual rate of such deaths per 100,000 people in the 1970s was less than one fifth of the 1960s level, while in the years 2010 to 2016 the annual rate of deaths from famine was just 1 per cent of the level of the 1960s. Taking a longer historic view, the current rate is barely a third of 1 per cent of the rate that was suffered in the 1870s. Even if we cast aside such rates, which take account of the growing world population, and look at absolute numbers of deaths, we find that more than twenty million people died of famine in the 1870s, and more than eighteen million perished in the 1940s; the figure for 2010 to 2016 is barely a quarter of a million.[35]

Yet at every stage of global population growth, Malthusians have worried that we are about to hit a new barrier that will lead to mass starvation. Indeed, such concerns have existed since the second century, when the theologian Tertullian warned that 'The final evidence of the fecundity of mankind is that we have grown too burdensome for the world: the elements scarcely suffice for our support, our needs grow more acute, our complaints more universal, since nature no longer provides us sustenance. In truth, pestilence and famine and war must be looked upon as a resolution for nations, a means of pruning the overgrowth of the human race.'[36]

As we have seen, similar concerns were expressed in the early twentieth century, before Haber's great breakthrough, and again in the 1960s, the time of peak global population growth. Paul Ehrlich's famous 1968 work *The Population Bomb* began arrestingly: 'The battle to feed humanity is over. In the 1970s the world will undergo famines – hundreds of millions of people are going to starve to death.'[37] In Ehrlich's defence, it might be argued that his alarmism motivated the programmes that caused a decline in population growth, but his underestimation of human innovation and its ability to produce more food has not chastened him in more recent times. In an interview to mark the fiftieth anniversary of his famous work in 2018, Ehrlich asserted that 'Population growth, along with over-consumption per capita, is driving civilization over the edge.'[38]

It is possible that this time it is different, but an apt analogy might be the joke about the man who, having jumped from the tenth floor of a building, passes the second floor and says, 'So far, so good.' There are two strands to such an argument. The first concerns environmental issues including global warming, while the second focuses on whether we can continue to increase food production at the rate required to feed the world, even if we have managed to do so in the past. I will return to the first of these questions in the Conclusion that follows this chapter, but now I want to ask whether we are finally running out of ideas to feed the world as we head towards a global population of ten or eleven billion.

There are some grounds for concern. In 2008, the World Development Report of the World Bank noted that increases in the yields of wheat, maize and rice in the developing world had been slowing since the 1980s. As the US environmentalist Lester Brown warned in 2005, 'diminishing returns are setting in on all fronts'.[39] However, measuring global food productivity is an extremely difficult task, and more recent work has suggested that the rise in the 'total factor productivity' of agriculture – what we get out relative to the land, labour and fertilizer that we put in – is, in fact, accelerating. One estimate is that the rate of total factor productivity growth doubled in the 1980s and 1990s, and continues to rise.[40] Part of the reason for this is the fact that fewer and fewer people are required to produce our food. In China, for example, the share of the workforce who are employed in the agricultural sector has fallen from more than half to less than a fifth in the last three decades.[41]

The data on total production is more certain than the data on productivity, yields or returns, since the calculation only requires the output data rather than being a ratio of output to input. Here the news is also reassuring: output growth has slightly accelerated in the twenty-first century. All the while, we must remind ourselves that population growth is slowing, which explains why the numbers of people suffering from malnutrition and famine are falling. There has been some slowing in output growth in the industrialized countries where food is plentiful anyway, but this has been offset by faster growth in the places where it is more needed.[42] It seems that global food supply is not only increasing but also that the developing world is becoming less dependent on the surpluses of developed nations; furthermore, there is a growing chance of export opportunities for poor countries, trade agreements permitting. With more investment in transport and refrigeration, agricultural wastage should also be reduced, which will lead to even more food being available to consumers.

Although climate change might reduce the productivity of some areas, it could increase the productivity of others, while technologies have been developed that make crops more resistant to heat.[43]

Another reason for optimism is that there continues to be a big gap between the world's agricultural leaders and the laggards. Indian cereal yields are well below half those achieved in the USA. In Jordan they are less than half those in Israel. And in Cuba they are only a little over half those in Brazil. No doubt there are area-specific reasons why not everywhere can reach the productivity of agriculture in the US Corn Belt, England's East Anglia or the Paris Basin, but there is surely room for some closing of the gap.

### *The Way We Live Now: Feeding the Eight Billion*

Rice is the staple diet of almost half the world's population.[44] About 90 per cent of all rice is grown in Asia, where China has the highest production and India the largest acreage. The Green Revolution, which has combined new strains with more fertilization, allowed rice yields to double in the last four decades of the twentieth century. Considering that what was state of the art in 1960 was the product of millennia of accumulated knowledge and experience, this doubling is a tribute to modern science.[45]

Just as the Haber–Bosch process was not the end point for improving yields, neither was the Green Revolution. Between 2000 and 2019, global rice paddy output grew by more than a quarter,[46] while Asian population expansion slowed to about 1 per cent a year during the same period. It is not surprising that since the start of the current century, the share of Chinese people who are underweight has halved, from around 16 per cent to around 8 per cent, while in East Asia as a whole it has fallen from 15 per cent to 5 per cent.[47] That is not bad progress in a region where the vast bulk of the population has been underfed for as long as agriculture has existed.

In the previous chapter, we noted the qualitative improvement in humans as they have become increasingly educated. But if education can be regarded as a software upgrade, having an adequately fed body is a hardware improvement. And as is so often the case with these sorts of developments, they feed off each other in a virtuous circle. The brain of a child who is well fed will develop better than that of a half-starved one, and a child who is not hungry can concentrate better at school. Equally, as we will see, a better-educated farmer is more likely to be productive, and better able to feed their family.

As ever, there are shadows to what is generally a happy story. First, not everywhere has improved as quickly as East Asia when it comes to food, and some places have gone backwards. For example, Zimbabwe, with its disastrous administration, has seen the share of its population who are underfed rise from 40 per cent to 50 per cent since 2000, scandalous in a country with excellent agricultural conditions and great potential. In war-torn Yemen, the share of the population who are underfed has sharply risen. Recent trends suggest that the overall number of hungry people in the world has started to rise, and the economic crisis and increase in poverty caused by the Covid-19 pandemic are likely, for a time, to make this worse.[48] With a growing population, a rise in the underfed can occur even as their share of the population decreases.

The poor are hardest hit when food prices rise, and this can lead to unrest such as the 'tortilla riots' in Mexico in 2007, India's onion crisis in 2013 and the demonstrations in response to the Egyptian government's cuts to bread subsidies in 2017. Fortunately for consumers, the FAO Food Price Index has been sharply lower since 2014, indicating greater general affordability. In real terms, however, the price of food is more or less where it was in the early 1960s, when the world had fewer than half as many mouths to feed.[49]

The other less happy part of the story concerns overeating. By 2007, the number of overweight people in the world outnumbered the hungry.[50] In some parts of the world, overeating is a pandemic that has serious implications for health and longevity. Humans have evolved to endure scarcity and gorge on food when it is available; many people find it difficult to control their appetites, and the US is famous for its problems in this regard. More than a third of its adults are obese, which as we saw earlier is contributing to a levelling off of life expectancy. Saudi Arabia has an obesity problem of similar proportions, and both countries also have high numbers of the merely overweight. In the Palestinian territories of the West Bank and Gaza, there are more than four overweight or obese children for every underweight one. While the overweight outnumber the underweight in both sexes, there are more obese boys than girls and more than twice as many underweight girls as boys. This reflects the priority many societies give to males when allocating resources.[51]

As populations urbanize, their diet changes – sometimes for the better. In modern cities, food safety standards are often higher than in the countryside, with better packaging, storage and refrigeration. But urbanites also eat more processed food, which means more sugar and salt, and leads to obesity, diabetes and increased blood pressure.

Globalization has been a large factor in the world's rising agricultural output. The USA been a major grain exporter since the nineteenth century, and its surpluses still feed much of the world, but the global food trade is about more than just US exports. For example, Brazil exports three-quarters of its soy crop to China, where it is used to feed animals and has underpinned the huge increase in Chinese meat-eating in recent decades. With globalization comes a reduction in self-reliance. Some people would see it as a cost, but the example of North Korea, which champions self-sufficiency in food production as well as in everything else, is hardly an advertisement for spurning such globalization: pre-school children there are up to thirteen centimetres shorter and seven kilograms lighter than their South Korean equivalents.[52]

## *A Farmer's Story*

Grain production in India has increased five-fold in sixty years, while its population has less than tripled. This is why, when I went to India in 2014 for the first time since the mid-1980s, people there seemed so much healthier and better-fed. Indeed, the experience of the last six decades in India can be captured in three simple multiples covering the last sixty years: less than three times the population, four times the yield and five times the output. Although more and more people can be fed from a given amount of land, India has performed less well than many other countries and has comparatively small farms and low yields.[53] This means that it still has scope for further improvement, while its population growth continues to fall steadily. The next time I travel to India, I expect to see even less hunger.

While aggregate statistics are all very well, it is worth understanding how they have been achieved. Unsurprisingly, the story has many parts: better irrigation, the use of better crop strains, upgraded farm equipment and increased access to fertilizers. Importantly, progress means the ability to use these more sparingly and effectively, leading to greater sustainability, but it turns out that education is one of the most powerful ways to enhance agricultural productivity.[54] A study of Indian rice farmers found a strong correlation between years of education and productivity whether or not modern techniques were adopted.[55]

A farmer called Chandranna took over his parents' three-acre smallholding in Karnataka, southern India. Although he did not attend university, he had some agricultural training in addition to his basic schooling. This led him to experiment with vermiculture to produce compost, which led to the highest yields of groundnuts in the area, with his bags weighing more than 50 per cent more than his neighbours'. This makes a real difference to his income and the daily life of his family; 'a modest mud house is now getting extended with cement walls,' a visitor has reported.[56] Many of Chandranna's neighbours are following his example; local stories like his are helping humanity out of the poverty that has endured since the dawn of time.

In some cases, the technology we now take for granted can make a real difference to people's livelihoods. For example, the mobile phone enhances agricultural productivity, providing an education tool, as well as delivering market information and access to micro-insurance. By 2016, there were apparently more mobile phones in Africa than toothbrushes.[57] The director of operations of a US non-profit organization that teaches farmers in western Kenya by text messaging told a *Financial Times* journalist that the technology helps by 'providing farmers with information and recommendations tailored to their local soil, weather and market conditions [that] could dramatically improve yields and net incomes'.[58]

Innovation continues, but its adoption is not always rapid. The lack of education or a resistance to change has long prevented advances in food production from spreading as quickly as they might.[59] Farms are sometimes too small for investment to be worthwhile, a problem that is getting worse in India, where the size of farms is declining.[60]

Genetically modified crops could provide higher yields and use less land, which would simultaneously reduce hunger and help wildlife. Each year, between a quarter of a million and half a million children in poor countries go blind from a lack of vitamin A, half of whom die within twelve months. Golden rice, a genetically modified strain that is widely available thanks to the biotechnology companies' waiving of patent rights, could prevent this.[61] The Gates Foundation believes that with improved fertilizer and the use of GM crops, African farmers could double their yields.[62] Other forms of 'biofortification', the engineering of crops to enhance their nutritional value, are already making a difference.[63]

The uptake of GM crops has been slower than might have been expected because of concerns about their health impact, the creation of super-weeds or the control they give to multinational corporations. Such worries, however, should be weighed against the very immediate needs of people to eat, and research does not generally support these concerns.[64] An approach that is both sustainable and can also feed the world is not beyond our reach.[65]

Progress in food production is necessary if we are to end world hunger, but famine, while it is generally related to food shortages, need not involve an absolute shortage.[66] It has often happened that while enough food is produced, it fails to reach the neediest recipients, so hunger persists. During the famines in Ireland, Ukraine and Bengal during the 1840s, 1930s and 1940s respectively, grain was still being exported. Food aid continues, although there is a concern that it distorts markets and disincentivizes local producers, and that it can be more about assisting farmers in the rich world than consumers in the poor. These are problems that go beyond the Malthusian idea that the world needs to produce enough food for all its mouths, however it might be distributed.

### *Foods of the Future*

Beyond the current model, some developments underway in food production could be completely transformative. Hydroponics is the system by which certain crops can be grown indoors rather than in soil, with a regulated environment and perfectly measured inputs, including LED light. One hydroponic facility, which calls itself a farm, is situated thirty-three metres below Clapham Common in south London and grows 20,000 kilograms of greens each year. Moving production underground frees up space elsewhere. All the produce is sold within the London area, removing the need to ship the food over long distances and meaning that it is extra-fresh when it reaches the consumer. 'We'll be cutting it at four in the afternoon and people will be eating it at the next lunchtime,' the co-founder Steven Dring told a reporter.[67] 'It's fantastic to source produce so fresh in the heart of Britain's largest city,' a well-known chef enthused.[68] Rooftop terraces that use hydroponics are emerging everywhere from Guangzhou to Montreal, and kits for the technique are even sold in Ikea.[69]

Plenty of other technologies that could transform how we produce food, the amount we can grow and the efficiency with which we farm it are still in their infancy; these could mean less artificial fertilizers and pesticides, less run-off, less use of land and greater sustainability. In the future, it may come to be seen as old-fashioned to indulge crops with vast quantities of space and uncontrolled inputs, in the hope that they will overcome the vagaries of nature.

Similarly, the day may not be so far off when breeding, husbanding and slaughtering an animal seems an absurd way of obtaining meat. Commercial lab-grown, or 'in-vitro', meat still seems a long way off, but the costs are falling fast.[70] A lab-grown burger that cost $280,000 as recently as 2013 may be produced for as little as $10 in the next few years.[71] The appeal of meat is widespread, but given its inefficiency as a way of obtaining food, some people see universal vegetarianism as environmentally and ethically preferable. Still, it may well be that rather than having to give up meat altogether, we can produce something very similar to it with less damage to the planet, less suffering to animals and potentially lower cost. The potential environmental upside is enormous; livestock grazing systems that produce meat and other animal products currently occupy more than a quarter of the globe's ice-free land surface and require vast amounts of resources.[72] Lab-created fish is also on the cards.

If endless exponential population growth were inevitable, we might have serious grounds for concern about whether human innovation would be able to cope. If the global population doubled four times every century and this continued for a number of centuries, we would surely run up against some kind of Malthusian frontier, whatever the level of human innovation. However, given that if we reduced wastage we could already feed a population of ten

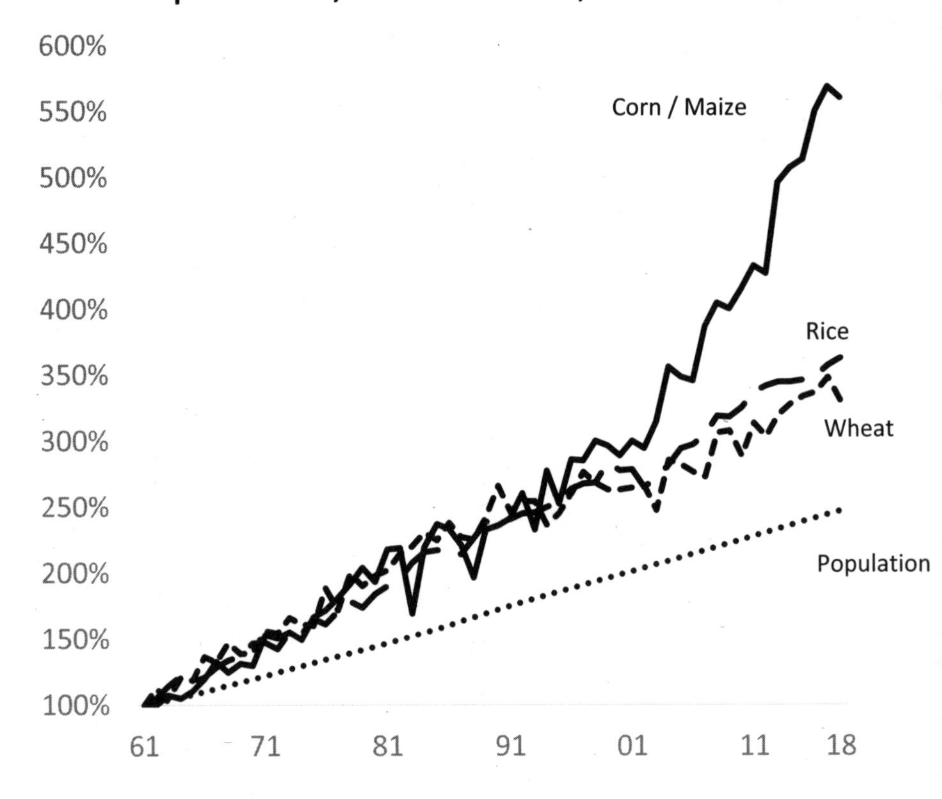


Source: FAO, UN Population Division

While population growth is now steady, growth in food production continues rapidly. Output of wheat and rice has more than tripled since the early 1960s, while population has risen two-and-a-half fold. However, the star performer in recent decades has been maize, the production of which has increased more than five-fold. This grain is predominantly used to feed farm animals for meat, which explains why it has been possible for per capita consumption to double over the last five decades.

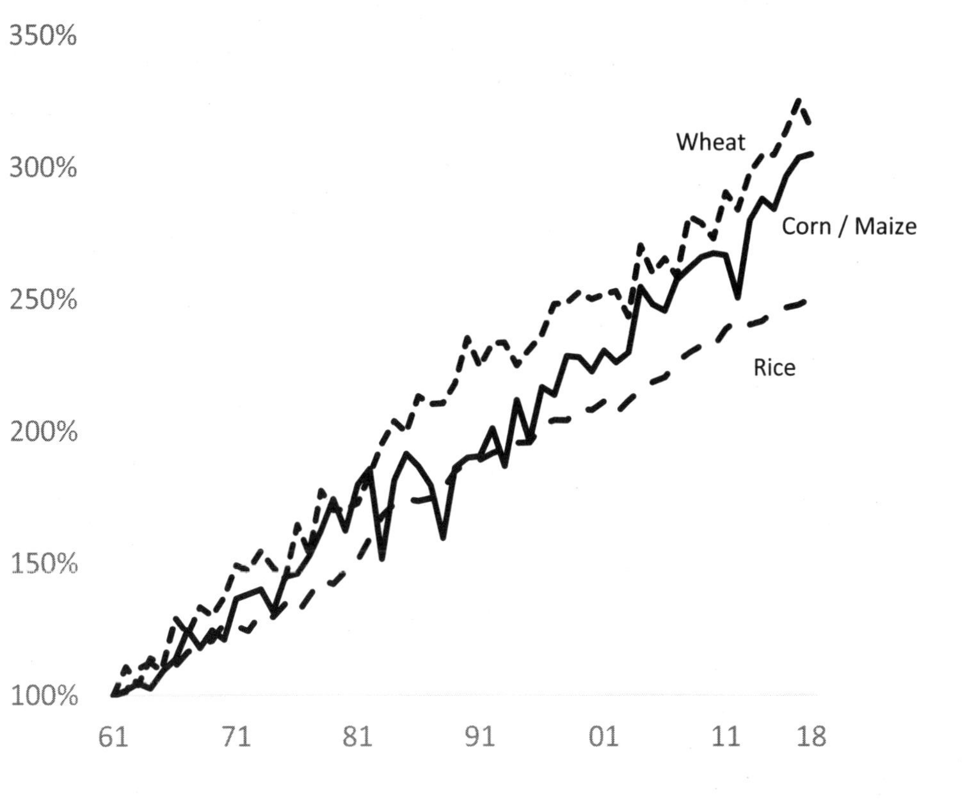


Source: FAO

Intensive farming has a bad reputation, but the production of more food from less land will allow more space to be given to nature. The recent great increase in crop yields has meant that we have not needed to devote vastly more space to agriculture in order to feed the world.

With population increases set to slow and much of the world focusing on eating less rather than more, a continued growth in yields will give a real chance for environmental recovery.

or eleven billion – the world's expected peak – and with plenty of innovation in the pipeline, there is every reason to believe that fewer people should be going hungry as the century progresses. Some people may resist innovations such as GM crops, but the poor will not have the luxury of doing so. And in any case, innovation in food production is hardly new; the move from hunting and gathering to agriculture involved a form of genetic engineering, albeit a gradual one. As the nineteenth-century US radical economist Henry George put it:

> Here is a difference between the animal and the man. Both the jayhawk and the man eat chickens, but the more jayhawks, the fewer chickens, while the more men, the more chickens. Both the seal and the man eat salmon, but when a seal takes a salmon there is a salmon the less, and were seals to increase past a certain point salmon must diminish; while by placing the spawn of the salmon under favourable conditions man can so increase the number of salmon as more than to make up for all he may take, and thus, no matter how much men may increase, their increase need never outrun the supply of salmon.[73]

Neither the factory farming of chickens nor the farming of salmon is free of environmental costs, but these can be managed. With an eye on future developments, Winston Churchill said: 'We shall escape the absurdity of growing a whole chicken in order to eat the breast or wing, by growing these parts separately under a suitable medium.'[74]

The ultimate barrier to population growth will not be food shortages or any other external factor, but the choices that humans make themselves.

# Conclusion

## *Tomorrow's People*

Tomorrow's people are emerging before our eyes. With life prolonged and death postponed, especially among the young, their overall numbers are booming, but the rate of increase is slowing as they produce fewer children. They are more urban, they are older and they are increasingly well-educated and well-nourished.

But as we have seen, the picture is far from uniform, for some parts of the world remain in the process of demographic modernization. While many countries in sub-Saharan Africa have already substantially reduced infant mortality and increased life expectancy, we can be fairly confident that the former will fall further while the latter will continue to rise. At the same time, fertility rates will keep sinking across Africa.

So for Africa, and for some countries beyond, like Afghanistan and East Timor, the future is about catching up with the rest of the world, as their populations grow bigger and older. We can expect the people who are currently on the bottom rung of the global ladder to become better fed and better educated. Although they will still be a long way from Denmark, that model of stability, prosperity and demographic maturity, the least prosperous parts of the world are heading in a Danish direction most quickly. It is highly likely that this sort of progress will continue, but four types of calamity could still prevent it: environmental disaster, war, pandemic and economic collapse. Let us now take a brief look at each of these.

This is not a book about global warming but it is inevitable that its shadow will fall over any discussion of the future. Rising greenhouse-gas emissions, temperatures and sea levels would lead to crop failures and a flood of climate refugees that would turn the assumptions in this book on their head. Even without such apocalyptic scenarios, we could see more pollution, less wildlife and the general devastation of life on our planet. No one can say for sure whether these things will occur, and some scientists believe that our situation is not nearly as bad as the alarmists would make out.[1] A lot of our activities require fewer emissions than they used to: an audio or video call with a loved one or a colleague can save a journey and consume far less energy, while modern LED bulbs consumes a fraction of the energy of old ones. Ultimately, development can be compatible, and even complement, protecting the environment.

With a larger and more educated global population, all sorts of innovations are being developed that will help the planet. Solar and wind power or carbon capture might solve the problem of emissions, and new ways of producing food might avoid the agricultural problems that are being caused by climate change. What definitely will *not* happen – at least on account of global warming – is the mass human extinction that some campaigners warn of. In fact, the number of people dying as a result of natural disasters has been falling consistently for decades, and such deaths currently amount to barely one in a thousand.[2] Quite apart from anything else, as people get richer, they are better able to protect themselves against calamities. Population growth will continue to put pressure on the environment, but this growth is slowing each year, while our encroachments on nature are already being reversed.

As with environmental disaster, war seems unlikely to have a significant future demographic impact. It always remains possible that

World War Three will break out tomorrow and decimate humanity, but the record is fairly clear. The proportion of people dying in battle has been on a generally falling trend and is a fraction of its level from the late 1960s.[3] The fact that at least 350,000 people have died in the decade of the Syrian conflict is, of course, a tragedy, but that number represents less than a year's population growth for the country before war broke out.[4] More demographically significant is the fact that ten times that number of people have left the country;[5] this is a loss of population for Syria, but millions of Syria's citizens now live in Jordan, Turkey, Lebanon or further afield. Furthermore, the number of deaths experienced in Syria over a decade is still tiny compared to those who died during just three years of war in Korea, for example.[6] Finally, as the global population ages, there is reason to hope that war will become less prevalent.

Much of this book has been written during lockdowns and isolation resulting from the coronavirus pandemic. Covid-19 has hugely dislocated the economy, and it may accelerate a move away from cities in the developed world. At the time of writing, an estimate of excess deaths because of the pandemic is over sixteen million.[7] Almost forty million people died of the Spanish flu a century ago, at a time when the world had less than a third of its current population.[8] The current annual global death toll from all causes is more than fifty million; since coronavirus disproportionately impacts the elderly, the Covid-19 pandemic will have a limited effect on life expectancy. If an event on the current scale of Covid-19 occurred once every century, the overall impact on our world's population would be negligible. As with the other disasters we have considered, it *could* end up being much worse and something comparable *might* arise before another hundred years has passed. But based on experience, there is no reason to fear that a disease will materially impact the number of people on the planet.

Covid-19 could have a more significant influence on births than deaths. Couples have been stuck at home during lockdowns, in many cases with relatively little to do, which could increase their sexual activity. The disruption to normal life has limited people's access to contraception, which the UN is concerned will lead to a boom in unplanned pregnancies.[9] On the other hand, several factors will tend to reduce the birth rate: a reluctance to go to hospital or attend medical appointments, the delay of weddings and the lack of opportunities for new sexual encounters, as well as serious economic uncertainty. People may well choose to delay having babies until they can control this. Overall, it is currently thought that the pandemic will depress fertility in the developed world, while the lack of access to contraception in less developed countries will increase births.[10] Whether the pandemic boosts or depresses childbearing, this is likely to be a short-term effect, although it will create 'lumpiness' for school and university admittances in the future.

A serious economic collapse, whether or not it is triggered by some other calamity, could be a body blow for poorer countries that are currently completing their demographic transition. In the economic crisis of 2008–9, the economies that were most damaged tended to be the poorest. Despite economic damage in much of the developing world in the wake of the 2008/9 economic crisis, infant mortality continued to fall and life expectancy continued to rise. It would take a much more seismic economic downturn to reverse these powerful forces.

### *The Day-After-Tomorrow's People*

We have so far focused on the tangible, with nothing more futuristic than lab-reared meat featuring in these pages. The trends in fertility, mortality, migration and ethnic change that will shape the future are clear, because they are already with us. However, it is time to consider some longer-term possibilities.

We have seen the beginnings of what could be a trend towards falling life expectancy in some of the world's most advanced countries. While any reversals are tentative, the extension of life expectancy is beginning to slow – even among the best performers. In Japan, life expectancy is currently growing by a year and a half or two years every decade, compared to five or six years every decade in the 1960s. It could be that human life expectancy is reaching some kind of natural limit. Diseases of despair and mass obesity could spread around the world, with both already particular problems of the Anglosphere.[11] Alcoholism, suicide and obesity have long been major problems in Russia.[12]

However, it is also possible that a significant scientific advance will affect our understanding of ageing and create a whole new vista. We need not think about *eternal* life; a life expectancy of 'just' two hundred years would radically change everything about the shape of society. When and how we study, our work patterns and our familial relationships would be quite different from those of today, even if we are unable to say exactly how. The life pattern many of us now take for granted might come to be seen as antiquated if we had twice as many years to live.[13]

In terms of our fertility, conception is already much less closely associated with sex. As contraception has improved, we have replaced chance with choice. It was never the case that all sex led to conception, but conception always resulted from sex, at least until the development of IVF technologies. In the future, the link between sex and conception could be severed entirely. Parents may not even come in pairs; individuals could choose to supplement their own genetic material with that not just of one other person, but with that of several others. Indeed, babies with the genetic material of more than two people have already been born.[14] The idea that the people of the future could be universally intelligent, beautiful and free of genetic diseases raises huge ethical questions, but the pressure to allow gene selection once it is technologically feasible will be immense. The fact that embryos are already negatively selected for certain diseases, with pregnancies terminated if they are detected, points the way.

With the possibility of babies having more than two parents, the basic categories of demography start to break down. Attributing each birth to a single mother and single father would become difficult, just as demographic boundaries that were once sharp are becoming blurred with the emergence of 'gender fluidity'. Transgender individuals currently account for a small share of the world's population, with 0.7 per cent – at most – identifying as transgender.[15] This might remain a phenomenon of a small minority, but it could also expand to the point where the majority select their gender from a spectrum and change it periodically. In such a world, considering the total fertility rate per woman or the life expectancy of males versus females would be meaningless.

Eventually, consciousnesses might be downloaded and reloaded into completely new bodies, or we could choose to live in a world of virtual reality. Artificial general intelligence might enslave, liberate or subsume human life, and we might establish settlements beyond the Earth. This may seem like the realm of science fiction, but much that we take for granted today would have been science fiction or magic to people a couple of hundred years ago. Once we are considering these kinds of possibilities, we are quite beyond the scope of this book.[16] But before we reach anything like two-hundred-year lives, designer babies, widespread gender change or downloaded consciousness, there are several more concrete items on the human agenda.

## *The Post-Modern Trilemma: The Three Es*

As the developed world heads into post-modern demography, there is a choice that can be described as a 'trilemma', a trade-off that is not between one of two options but two of three. These options can be called 'the three Es': **economics**, by which I mean the buoyant economic growth we think of as 'normal'; **ethnicity**, the continued predominance of a single particular ethnic group within the territory it regards as its homeland; and **egoism**, specifically the prioritizing of personal projects above family formation.

I am using 'egoism' here as shorthand for something more complicated. The reason that people put off having children and end up having either few or none – except, of course, when they are not able to have children – is often a response to pressures around work, financial constraints, the requirement to care for aged parents and all sorts of social pressures as well as personal desires.[17] No choices are made in a vacuum. Women often face the toughest burdens around combining paid work with caring for their family and household chores, and they also tend to face the most pressure to bear the next generation. Financial or other constraints prevent people having children, and in most of the developed world they would like to have more than they do.[18] The term 'egoism' should be seen as summing up the pressures and preferences that result in people choosing to have small families or no families at all.

This trilemma can best be illustrated by three countries that have each traded off one of the options in order to enjoy the other two. Let's start with Japan, where ethnicity and egoism have been preserved at the expense of the economy. As we have seen, the Japanese have not been prepared to open their country to large-scale immigration. Multi-culturalism is not something most Japanese people would welcome,[19] but nor are they choosing to have children. They are not helped by a culture that discourages women from combining work and parenthood, while expecting them to undertake the vast majority of the housework and care – in such a situation, it is not surprising that many prefer independence. The Japanese have sacrificed dynamic economic growth, while racking up government debt at a world-beating rate. A declining working-age population, followed by a drop in the size of the population as a whole, is acting as a huge drag on economic growth that is unlikely to be repaired by any economic intervention.

There is much debate about whether immigration is beneficial to an economy, both in the short term and when measured by per capita income.[20] It may be that it does not assist the income of workers who are already in the country, but without population growth, an economy will struggle to grow. Within a country, there may be winners and losers from immigration. The UK, for example, surely has a larger economy in total – if not necessarily per capita – because of its nine million foreign-born people (around 13 per cent of the population). A decline in the number of workers is a drag on economic growth, while having more workers boosts it. And without the extra labour that immigration supplies, shortages will eventually appear in the labour force of countries that have relatively few native-born entrants into the workforce, following years of low fertility.

Individuals may be most concerned about their own income, but they will feel the impact of labour shortages when there are not enough teachers at their child's school or when there are too few nurses or carers to look after an elderly parent. Governments, on the other hand, are most concerned about the overall size of their economies, annual GDP growth and tax income, as well as the availability of workers needed to keep the economy running smoothly and maintain the availability of services.

The UK has taken a different path to Japan's. Its citizens are also relatively unenthusiastic about having children; UK fertility rates have been below replacement rate for about fifty years now, never mind at the levels required to create the steadily growing workforce required for a buoyant economy in the absence of high productivity growth. Instead, the UK has opted for mass immigration to plug the gap.

Britain has retained at least some of its economic dynamism, and has been better able than Japan to staff its hospitals, schools and offices, although the ethnic composition of the country has changed. White people of British origin made up well over 90 per cent of the population as recently as the 1990s, but barely 80 per cent by 2011; their share will decline further in coming decades. Some people believe that rapid ethnic change is not a trade-off but a benefit to be welcomed. In most societies, however, the idea of the traditional majority becoming a minority in what they regard as their homeland is met with some resistance. While minorities may assimilate into the majority group, their fertility rate then falls to that of society as a whole, leaving the long-term demographic problem unsolved.

The persistence of a nation or ethnic group is not guaranteed. Just as there are no more Medes or Visigoths, there is no guarantee that there will, in the future, be any Italians or Japanese The UN expects both countries to lose more than a third of their populations in just the next eighty years.

There are other examples of developed countries like Japan that have chosen to sacrifice economic growth for ethno-continuity and the societal egoism that discourages procreation. There are also plenty of examples of countries like the UK, which are undergoing fast ethnic change in order to maintain economic growth and where family formation is not a priority. But Israel is the only example of a country where a pro-natal culture has led fertility rates to rise to the point where the average woman is having three children. In fact, no other developed country is close to this. Israel was built on the immigration of Jews, but now it is high fertility that is ensuring its Jewish majority. Jewish women in Israel currently have a slightly higher fertility rate than Israel's Arab citizens.[21] (Note however that Palestinians in Gaza still have a somewhat higher, although greatly reduced, fertility rate.) They have preserved economic dynamism and ethnic continuity, while sacrificing the egoism of a childless life.

Again, 'egoism' should be seen here as including social pressures that work against childbearing; in Israel's case, not only are people prepared to make the trade-offs in order to have children, but they are also responding to social pressures that strongly encourage large families, from government policy and the provision of services, to an intangible but pronounced environment in which those people who don't have children feel the cold shoulder of disapproval.

Israel may be a special case, having been surrounded since its foundation by hostile neighbours and with its significant religious population, but there is no reason why people in Iceland or Italy should not have just as many children – after all, it is about culture rather than economics, priority rather than biology. This is the essence of post-modern demography. Israel has shown that a modern country can have high levels of education and long life expectancy while preserving a child-oriented and pro-natal culture.

The above might be read as a call for pro-natalism. It is certainly true that parenthood has been the most fulfilling thing I have done, but my goal here is not to preach but to point out the demographic options that individuals and countries face – along with their consequences. The ability of governments to help people plan their families when they are at an early stage of economic development, and therefore likely to want to *reduce* their fertility, is well proven. The savagery of India's enforced sterilizations of the 1970s was unnecessary, and so was the coercion in China, but it is much harder for governments to *increase* the birth rate. Tax and welfare incentives may help slightly, as do laws and subsidized childcare that help its women combine work and motherhood. However, in the post-modern world, what matters most are the preferences and actions of individuals and families.

Some people on the left spurn any kind of pro-natalism and any pressure that discourages people from pursuing their personal goals. However, they must ask themselves whether their individualistic, secular societies can survive in the face of persistent low fertility. Wealthy countries can draw in immigrants for now, but they will either retain their traditional values and undermine the secular, progressive values that liberals cling to, or they will assimilate, adopting low fertility themselves and failing to provide a long-term demographic answer. Those people on the right who lament the waning of the predominance of their ethnic or national group in their homelands must ask themselves whether they can legitimately complain about being 'replaced' if they are not prepared to reproduce themselves. So in the end, the destiny of tomorrow's people will depend more than anything on the choices made by the people of today.

## *Acknowledgements*

Professor Eric Kaufmann has been unflagging in his years of encouragement of my writing and has made many valuable comments on the manuscript. Professor Danny Dorling has been extraordinarily generous with his time and thoughts. I am indebted too to David Goodhart for his comments on the text and to old friends Robert Marshall, Ian Price and Michael Wegier. Martin van der Weyer was kind enough to comment on some of the economics-related topics. I have enjoyed many stimulating discussions on demography with Richard Ehrmann. My views on several of the subjects covered have emerged from many years of discussion with Nick Lowcock. I have appreciated the unique environment provided at St Antony's College, University of Oxford, by Professor Roger Goodman and his colleagues.

Michele Rosen was a great help on the manuscript, as was Nicholas Humphrey. The book has greatly benefited from the meticulous attention of Nicholas Blake. And I cannot too highly sing the praises of Toby Mundy who is so much more than an agent and whose intelligence, curiosity and professionalism have been invaluable at every stage.

Nothing I do would be possible without the love and support of my wife Claire, of my children and children-in-law Sonia and Joel, Juliet and Samuel and Adam, and of my mother Ingrid Morland, to whom this book is dedicated.

# *Notes*

## INTRODUCTION

1 See Morland, Paul, *The Human Tide: How Population Shaped the Modern World*, London, John Murray, 2019.
2 Livi-Bacci, Massimo, *The Population of Europe*, Oxford, Blackwell, 2000, p. 120.
3 Livi-Bacci, Massimo, *A Concise History of World Population*, Chichester, Wiley-Blackwell, 2012, pp. 41–3.
4 For a discussion of how uneconomic and impractical it was to transport food as late as the eighteenth and early nineteenth centuries, see Blanning, Tim, *The Pursuit of Glory: Europe 1648–1815*, New York, Viking, 2007, pp. 3–34.
5 Wilson, Peter H., *Europe's Tragedy: A New History of the Thirty Years War*, London, Penguin, 2010, p. 787; Lee, Harry F. and Zhang, David D., 'A Tale of Two Population Crises in Recent Chinese History', *Climatic Change*, 116, 2013, pp. 285–308.
6 This overlooks the Renaissance, which can be seen as spanning the medieval and modern. These periodizations are inherently imperfect and open to challenge.
7 Later editions of Malthus's *Essay* allowed for a greater tendency of people in some places and at some times to hold back their numbers below the maximum which resources could allow and thereby to improve their standard of living beyond subsistence.
8 For a masterly historiography of demographic transition theory, see Kirk, Dudley, 'Demographic Transition Theory', *Population Studies*, 50 (3), 1996, pp. 361–87.
9 For a discussion of the relative impact of economics, culture, institutions and other factors in driving the demographic transition see Kirk, op. cit., passim.

10 These demographic data, like all those in this book which are not endnoted, come from the United Nations Population Division. The income data are from the World Bank – GNI Atlas Method: https://data.worldbank.org/indicator/NY.GNP.PCAP.CD?view=chart (impression: 30 September 2020). The literacy data are also from the World Bank: https://data.worldbank.org/indicator/SE.ADT.LITR.FE.ZS?locations=MA (impression: 30 September 2020).

11 I have quantified this by correlating GPD per capita with fertility, life expectancy and infant mortality in 1970 and 2019 for more than one hundred countries for which the relevant data is available. I have found that the correlation between income and each of these demographic measures (a positive correlation for the first two, a negative one for the third) weakened between these two dates. The correlation between GDP per capita and fertility has weakened significantly more than that between life expectancy or (negatively) infant mortality. And the correlation in 2019 is much weaker for richer countries than for poor ones.

12 Most notably Gordon, Robert, *The Rise and Fall of American Growth: The U.S. Standard of Living Since the Civil War*, Princeton, New Jersey, Princeton University Press, 2016.

13 Kaa, D. J., van de, *Europe's Second Demographic Transition*, Washington DC, Population Reference Bureau, 1987; Lesthaeghe, R., *The Second Demographic Transition in Western Countries: An Interpretation*, Brussels, Interuniversity Programme in Demography, 1991; Lesthaeghe, R., 'The Unfolding Story of the Second Demographic Transition', *Population and Development Review*, 36 (2), 2010, pp. 211–51; see also Ariès, Philippe, 'Two Successive Motivations for the Declining Birth Rate in the West', *Population and Development Review*, 6 (4), 1980, pp. 645–50.

14 Kaa, Dirk J. van de, 'Europe's Second Demographic Transition', *Population Reference Bureau*, 42 (1), 1987, p. 46.

15 Lesthaeghe, R., 'The Second Demographic Transition: A Concise Overview of its Development', *PNAS*, 111 (51), 2014

16 For a fuller discussion of this, see Morland, op. cit., pp. 29–33, 283–90.

17 Drixler, Fay, *Infanticide and Population Growth in Eastern Japan 1660–1950*, Berkeley, University of California Press, 2013, pp. 18–19, 33, 124.

18 United Nations Population Division: https://population.un.org/wpp/Download/Standard/Population/ (impression: 2 October 2020).

## CHAPTER I: INFANT MORTALITY

1 World Bank: https://data.worldbank.org/indicator/SP.DYN.IMRT.IN?locations=PE (impression: 19 July 2021).
2 *Anglican Journal*, 6 November 2019: https://www.anglicanjournal.com/indigenous-midwives-exchange-knowledge-pwrdf-program-shares-best-practices-from-canada-mexico-and-peru (impression: 21 November 2019).
3 World Bank: https://data.worldbank.org/indicator/SP.DYN.IMRT.IN?locations=PE (impression: 27 July 2020).
4 World Bank: https://data.worldbank.org/indicator/SE.SEC.ENRR.FE?locations=PE (impression: 6 March 2019).
5 Kiross, Girmay Tsegay et al., 'The Effect of Maternal Education on Infant Mortality in Ethiopia: A Systematic Review and Meta-analysis', *PLoS One*, 14 (7) 2019.
6 Case, Anne and Deaton, Angus, *Deaths of Despair and the Future of Capitalism*, Princeton and Oxford, Princeton University Press, 2020, pp. 57, 59, 66, 75–7.
7 *Peruvian Times*, 30 December 2013: https://www.peruviantimes.com/30/perus-maternal-mortality-rate-down-50-in-last-10-years/21077/50-in-last-10-years/21077/ (impression: 18 December 2018).
8 Minello, Alessandra, Dalla-Zuanna, Gianpiero and Alfani, Guido, 'First Signs of Transition: The Parallel Decline of Early Baptism and Early Mortality in the Province of Padua (north-eastern Italy), 1816–1870', *Demographic Research*, 36 (1), 2017, p. 761. Note that while the matching trends are clear, the authors are open to different interpretations of the data.
9 Henry's last wife, Catherine Parr, had one child by her fourth husband after Henry VIII's death. This child died around the age of two.
10 It is true that there is a counter-tendency to embrace death. Bach's exceptionally beautiful Cantata BWV 82 commences with a lament, 'Ich habe genug' (I have had enough), and ends with 'Ich freue mich auf meinen Tod' (I rejoice in my death). And apart from artists, mystical pietists and the odd psychopath, there have always been individuals who for one reason or other have embraced or even precipitated their own death such as those seeking glory, martyrdom and the possibility of eternity. A fervent belief in a better afterlife or in reincarnation may reduce the desire to stay alive. But had such cases not been exceptional, the human race could hardly have survived and thrived. The will to life is generally the stronger emotion.

11 I thought the use of this analogy was original. I then read it in Case and Deaton, op. cit., p. 22.
12 Alberts, Susan C., 'Social Influences on Survival and Reproduction: Insights from a Long-Term Study of Wild Baboons', *Journal of Animal Ecology*, 23 July 2018, p. 50: https://besjournals.onlinelibrary.wiley.com/doi/10.1111/1365-2656.12887 (impression: 15th November 2019)
13 *Independent*, 28 July 2013: https://www.independent.co.uk/news/world/politics/220-million-children-who-dont-exist-a-birth-certificate-is-a-passport-to-a-better-life-so-why-cant-8735046.html (impression: 22 November 2019).
14 Bricker, Darrell and Ibbitson, John, *Empty Planet: The Shock of Population Decline*, New York, Crown, 2019, p. 68.
15 UPI, 16 July 2020: https://www.upi.com/Health_News/2020/07/16/US-infant-mortality-rate-hits-all-time-low-CDC-reports/8081594905861/ (impression: 27 July 2020).
16 Center for Disease and Control and Prevention: https://www.cdc.gov/reproductivehealth/maternalinfanthealth/infantmortality.htm (impression: 27 July 2020).
17 Department of Health and Human Services: https://minorityhealth.hhs.gov/omh/browse.aspx?lvl=4&lvlid=68 (impression: 27 July 2020).
18 WCPO Cincinatti: https://www.wcpo.com/news/transportation-development/move-up-cincinnati/cradle-cincinnati-2018-infant-mortality-rate-improves-but-remains-far-higher-for-black-babies (impression: 27 July 2020).
19 Center for Diseases Control and Prevention: https://www.cdc.gov/nchs/pressroom/sosmap/infant_mortality_rates/infant_mortality.htm (impression: 27 July 2020).
20 http://www.newindianexpress.com/world/2019/feb/04/maldives-indian-coast-guard-successfully-evacuates-critically-ill-infant-1934136 (impression: 21 August 2019).
21 ONS: https://www.ons.gov.uk/peoplepopulationandcommunity/birthsdeathsandmarriages/deaths/bulletins/childhoodinfantandperinatalmortalityinenglandandwales/2018 (impression: 27 July 2020).
22 *Guardian*, 19 April 2019: https://www.theguardian.com/society/2019/apr/19/newborn-baby-deaths-may-be-on-rise-among-poorest-in-england (impression: 24 January 2020).
23 ONS: https://www.ons.gov.uk/peoplepopulationandcommunity/birthsdeathsandmarriages/deaths/bulletins/childhoodinfantandperinatalmortalityinenglandandwales/2019 (impression: 19 July 2021).

24 *Guardian*, 6 December 2019: https://www.theguardian.com/lifeandstyle/2019/dec/06/record-number-of-over-45s-giving-birth-in-england (impression: 23 November 2020).
25 UNICEF: https://data.unicef.org/resources/levels-and-trends-in-child-mortality/ (impression: 27 July 2020).
26 Cato Institute, 3 April 2019: https://www.cato.org/publications/commentary/human-progress-saved-baby-will-save-many-more (impression: 27 July 2020).
27 Ibid.
28 Agence France Press, 10 February 2014: https://www.pri.org/stories/2014–02–10/pakistan-where-conspiracy-theories-can-cost-childs-life (impression: 7 July 2020).
29 *Daily Telegraph*, 29 April 2019: https://www.telegraph.co.uk/global-health/science-and-disease/pakistan-polio-vaccinations-halted-killings-amid-panic-sterilization/ (impression: 27 July 2020).
30 Ntedna, Peter Austin Morten and Tiruneh, Fentanesh Nibret, 'Factors Associated with Infant Mortality in Malawi', *Journal of Experimental and Clinical Medicine*, 6 (4), August 2014, pp. 125–9: https://www.researchgate.net/publication/263812940_Factors_Associated_with_Infant_Mortality_in_Malawi (impression: 13 December 2019).
31 Full Fact: https://fullfact.org/health/how-many-people-die-fires/ (impression: 13 December 2020).
32 Interview with Michael Rosato, CEO of Women and Children First, 12 December 2019, London.
33 Ibid.
34 *Guardian*, 19 September 2019: https://www.theguardian.com/global-development/2019/sep/19/number-women-dying-childbirth-off-track (impression: 22 January 2019).
35 Al Jazeera, 14 March 2016: https://www.aljazeera.com/indepth/features/2016/03/sri-lanka-beats-india-maternal-mortality-ratios-160308105127735.html (impression 4 February 2019).
36 World Bank: https://data.worldbank.org/indicator/SH.STA.MMRT?locations=LK (impression: 27 July 2020).
37 Data 1990–2013. Regions halving or more in this period include Northern Africa and East Asia. Trends in Maternal Mortality 1990-2013 WHO et al, p. 25: http://apps.who.int/iris/bitstream/handle/10665/112682/9789241507226_eng.pdf;jsessionid=C8C8E09C1B10F77323BE5B0C992A4250?sequence=2 (impression: 21 October 2021).

38 Prost, Audrey et al., 'Women's Groups Practicing Participatory Learning and Action to Improve Maternal and New Born Health in Maternal Settings: A Systematic Review and Meta-Analysis', *The Lancet*, 2013, 381, 1736–46.
39 UNICEF: https://data.unicef.org/topic/maternal-health/maternal-mortality/ (impression: 27 June 2020); *Guardian*, 30 January 2017: https://www.theguardian.com/global-development/2017/jan/30/maternal-death-rates-in-afghanistan-may-be-worse-than-previously-thought#img-1 (impression: 12 October 2019). Note, however, that other sources put it lower e.g. the World Bank: https://data.worldbank.org/indicator/SH.STA.MMRT?locations=AF (impression: 19 July 2021).
40 *Journal of the Royal Society of Medicine*, November 1999: https://www.ncbi.nlm.nih.gov/pmc/articles/PMC1633559/ (impression: 4 February 2019 ); World Bank, op. cit.
41 Other data puts the rate higher, at more than twenty-six deaths per hundred thousand pregnant women, but this is taking account of deaths in pregnancy rather than after birthing: https://www.health.harvard.edu/blog/a-soaring-maternal-mortality-rate-what-does-it-mean-for-you-2018101614914 (impression: 9 September 2020).
42 CNN, 20 February 2018: https://edition.cnn.com/2018/02/20/opinions/protect-mother-pregnancy-williams-opinion/index.html (impression: 17 August 2020); Independent, 9 December 2019, https://www.independent.co.uk/life-style/women/beyonce-miscarriage-pregnancy-loss-life-lessons-blue-ivy-jay-z-elle-uk-a9239121.html (impression: 21 October 2021).
43 *New York Times*, 7 May 2019: https://www.nytimes.com/2019/05/07/health/pregnancy-deaths-.html (impression: 13 December 2019).
44 Centers for Disease and Control Prevention: https://www.cdc.gov/vitalsigns/maternal-deaths/index.html (impression: 13 December 2019).

## CHAPTER 2: POPULATION GROWTH

1 The UN median forecast for Africa as a whole is somewhat higher, for sub-Saharan Africa somewhat lower. In general, unless otherwise stated, the data mentioned in this chapter for Africa is for sub-Saharan Africa only.
2 *Financial Times*, 17 November 29016: https://www.ft.com/content/8411d970-7b44-11e6-ae24-f193b105145e (impression: 21 August 2019).
3 Ibid.

4 *Africa Times*, 27 November 2019: https://africatimes.com/2019/11/27/iom-climate-change-a-clear-driver-of-african-migration/ (impression: 29 November 2019).
5 *Climate Home News*, 16 May 2019: https://www.climatechangenews.com/2019/05/16/lake-chad-not-shrinking-climate-fuelling-terror-groups-report/ (impression: 29 November 2019); BBC 27 September 2018: https://www.bbc.co.uk/news/world-africa-45599262 (impression: 29 November). *Guardian*, 22 October 2019: https://www.theguardian.com/global-development/2019/oct/22/lake-chad-shrinking-story-masks-serious-failures-of-governance (impression: 24 January 2019).
6 The accuracy of this statement will depend on how soon after its being written it is read, as is the case with all fast-changing phenomena. By mid-century, the UN's median forecast is that Niger's population will be as large as the UK's is today.
7 Euronews, 31 October 2019: https://www.euronews.com/2019/10/31/being-a-malnourished-child-in-niger-two-stories (impression: 29 November 2019).
8 Segal, Ronald, *Islam's Black Slaves: The History of Africa's Other Diaspora*, London, Atlantic Books, 2001, pp. 56–7.
9 Elton, J. Frederic, *Travels and Researches among the Lakes and Mountains of Eastern and Central Africa*, London, John Murray, 1879, p. 23.
10 Deutscher, Guy, *Through the Language Glass: How Words Colour Your World*, London, William Heinemann, 2010, p. 164.
11 Darwin, Charles, *The Descent of Man and Selection in Relation to Sex*, New York, D. Appleton and Company, 1871, p. 193.
12 Horsman, Reginald, *Race and Manifest Destiny: The Origins of American Racial Anglo-Saxonism*, Cambridge MA and London, Harvard University Press, 1981, pp. 243–4.
13 *Washington Spectator*, 2 November 2019: https://washingtonspectator.org/italy-and-beyond/ (impression: 1 December 2019).
14 BBC News, 14 November 2019: https://www.bbc.co.uk/news/stories-50391297 (impression: 1 December 2019).
15 Smith, Stephen, *The Scramble for Europe: Young Africa on its Way to the Old Continent*, Polity, Cambridge, 2019, p. 159.
16 Collier, Paul, *Exodus: Immigration and Multiculturalism in the 21st Century*, Penguin, London, 2014, pp. 41–3.
17 Reuters, 15 August 2020, https://www.reuters.com/article/us-italy-migrants-minister-idUSKCN25B0SO (impression: 22 October 2021).

18 *Mail & Guardian*, http://atavist.mg.co.za/ghana-must-go-the-ugly-history-of-africas-most-famous-bag (impression: 24 January 2019).
19 All Africa, 28 February 2018: https://allafrica.com/stories/201803010011.html (impression: 24 January 2018).
20 Migration Data Portal https://www.migrationdataportal.org/regional-data-overview/southern-africa (impression: 21 October 2021).
21 Brookings, 7 June 2018: https://www.brookings.edu/blog/africa-in-focus/2018/06/07/figures-of-the-week-internal-migration-in-africa/ (impression: 24 January 2019).
22 Quartz, 28 March 2019: https://qz.com/africa/1582771/african-migrants-more-likely-to-move-in-africa-not-us-europe/ (impression: 24 January 2019).
23 Fukuyama, Francis, *Political Order and Political Decay*, Profile, London, 2014, pp. 25–7.
24 Vollset, Stein Emil et al., 'Fertility, Mortality, Migration and Population Scenarios for 195 Countries and Territories: A Forecasting Analysis for the Global Burden of Disease Study', *The Lancet*, 14 July 2014.
25 Kaufman, Carol E., 'Contraceptive Use in South Africa Under Apartheid', *Demography*, 35 (4), 1998, pp. 421–34.
26 All Africa, 21 November 2019: https://allafrica.com/stories/201911270852.html (impression: 6 December 2019).
27 New Security Beat, 11 May 2015: https://www.newsecuritybeat.org/2015/05/whats-west-central-africas-youthful-demographics-high-desired-family-size/ (impression: 6 December 2019).
28 Devex, 21 November 2019: https://www.devex.com/news/innovative-approaches-to-improving-contraceptive-access-in-kenya-96048 (impression: 16 December 2019).
29 Knoema: https://knoema.com/atlas/Nigeria/topics/Education/Literacy/Adult-femaleilliteracy (impression: 26 January 2020).
30 Livi-Bacci, Massimo, *The Population of Europe*, Oxford, Blackwell, 1999, p. 165.
31 One, 28 November 2018: https://www.one.org/international/blog/aids-facts-epidemic/?gclid=EAIaIQobChMI_7Orgt2c5wIVQrTtCh2-DA9yEAAYASAAEgLC7_D_BwE (impression: 24 January 2020).
32 Avert: https://www.avert.org/professionals/hiv-around-world/sub-saharan-africa/swaziland (impression: 21 October 2021).
33 AAAS Science Magazine, 24 July 2017: https://www.sciencemag.org/news/2017/07/swaziland-makes-major-strides-against-its-aids-epidemic (impression: 8 December 2019).

34 Center for Disease Control and Prevention: 2014–2016 Ebola Outbreak in West Africa: https://www.cdc.gov/vhf/ebola/history/2014-2016-outbreak/index.html (impression: 15 December 2019).
35 World Bank: https://data.worldbank.org/indicator/NY.GDP.PCAP.PP.CD (impression: 20 March 2020).
36 *Financial Times*, 22 November 2019: https://www.ft.com/content/69f907ce-e127-11e9-b8e0-026e07cbe5b4 (impression: 15 December 2019).
37 Whether Europe's wars really have been about resources is of course highly debatable. The Lenin / Imperialism view of World War One is that they were – but that view has largely been discredited. Nevertheless, the desire for territory, whether Britain versus Germany in the colonies or Austrian ambitions against Russian clients in the Balkans, was often not just about glory or self-aggrandisement or power but also about taxable and extractable resources. The Franco-German bitterness over Alsace-Lorraine for example was intensified because of the significant iron deposits to be found there.

## CHAPTER 3: URBANIZATION

1 United Nations, The World's Cities in 2018: https://www.un.org/en/events/citiesday/assets/pdf/the_worlds_cities_in_2018_data_booklet.pdf (impression: 23 October 2020). This includes Hong Kong but not the cities of Taiwan.
2 Researchgate: https://www.researchgate.net/figure/Worlds-largest-cities-megacities-in-the-world-1900-2015_tbl1_226618338 (impression: 21 August 2019).
3 *Guardian*, 20 March 2017: https://www.theguardian.com/cities/2017/mar/20/china-100-cities-populations-bigger-liverpool (impression: 21 August 2019).
4 World Population Review: https://worldpopulationreview.com/countries/cities/india (impression: 19 August 2020).
5 Luxembourg – A Small but Open Society: https://luxembourg.public.lu/en/society-and-culture/population/demographics.html (impression: 19 August 2020).
6 Childe, V. Gordon, 'The Urban Revolution', *Town Planning Review*, 21 (1) 1950, pp. 3–17.
7 China Today, 2 November 2018 http://www.chinatoday.com.cn/ctenglish/2018/tourism/201811/t20181102_800146032.html (impression: 21 October 2021).

8 Slate, 31 October 2013: https://slate.com/news-and-politics/2013/10/nanchang-china-a-city-the-size-of-chicago-that-youve-never-heard-of.html (impression: 21 August 2019); MarcroTrends: https://www.macrotrends.net/cities/20622/nanchang/population (impression: 21 August 2019).
9 UN Population Division: https://population.un.org/wup/Archive/Files/studies/United%20Nations%20(1977)%20-%20Orders%20of%20magnitude%20of%20the%20world%27s%20urban%20population%20in%20history.PDF; World Bank: https://data.worldbank.org/indicator/sp.urb.totl.in.zs; UN: https://www.un.org/development/desa/en/news/population/2018-revision-of-world-urbanization-prospects.html (impressions: 21 August 2019).
10 World Population Review: http://worldpopulationreview.com/world-cities/chongqing-population/ (impression: 2 February 2020).
11 Knoema: https://knoema.com/atlas/China/Urban-population (impression: 19 August 2020).
12 Statistica.com: https://www.statista.com/statistics/289158/telephone-presence-in-households-in-the-uk/ (impression: 14 February 2020).
13 Crosby, Alfred W., *The Measure of Reality: Quantification and Western Society, 1250–1600*, Cambridge, Cambridge University Press, 1997, p. 129.
14 Evans, Richard J., *The Pursuit of Power: Europe 1815–1914*, London, Allen Lane, 2016, p. 8.
15 Shan, Weijian, *Out of the Gobi: My Story of China and America*, Hoboken, New Jersey, Wiley, 2019, p. 135.
16 Twine, Kevin, 'The City in Decline: Rome in Late Antiquity', *Middle States Geographer*, 25, 1992, p. 136.
17 Marsden, Peter and West, Barbara, 'Population Change in Roman London', *Britannia*, 23, 1992, pp. 133–40.
18 Davis, Kingsley, 'The Urbanization of the Human Population', in LeGates, Richard T. and Stout, Frederic, eds., *The City Reader*, London and New York, Routledge, 2016, p. 481.
19 Morland, Paul, *The Human Tide: How Population Shaped the Modern World*, London, John Murray, 2020.
20 Again, a lot depends on precisely how you define urban, which requires both a determination of the minimum the size of a qualifying conurbation and the conurbation's boundaries.
21 *Financial Times*, 24 March 2020: https://www.ft.com/content/1df725c0-6adb-11ea-800d-da70cff6e4d3 (impression: 24 March 2020).

22 *Guardian*, 23 March 2009: https://www.theguardian.com/environment/2009/mar/23/city-dwellers-smaller-carbon-footprints (impression: 16 February 2020).
23 Live Science, 19 April 2011: https://www.livescience.com/13772-city-slicker-country-bumpkin-smaller-carbon-footprint.html (impression: 16 February 2020).
24 US Energy Information Administration https://www.eia.gov/environment/emissions/state/analysis/ (impression: 21 October 2021).
25 Peter Calthorpe cited in Brand, Stewart, *Whole Earth Discipline: An Ecopragmatist Manifesto*, London, Atlantic Books, 2010, p. 67.
26 Ibid., p. 68.
27 Smil, Vaclav, *Growth: From Microorganisms to Megacities*, Cambridge, Mass. and London, MIT, 2019, p. 343.
28 Pyrenean Way: http://www.pyreneanway.com/2014/06/rewilding-and-the-pyrenees/?lang=en; The Connexion: https://www.connexionfrance.com/French-news/Camera-captures-rare-brown-Pyrenees-bear (impressions: 14 February 2020).
29 Flyn, Cal, *Islands of Abandonment*, London, William Collins, 2021, pp. 53, 59.
30 Wired, 14 February 2018: https://www.wired.co.uk/article/tfl-finances-transport-for-london-deficit-passenger-numbers (impression: 14 February 2020).
31 Carter, Mike, 'Stranded in Paradise: A Spring Awakening amid the Welsh Hills', *Financial Times*, 5 May 2020: https://www.ft.com/content/f095f452-8309-11ea-b6e9-a94cffd1d9bf (impression: 18 August 2020).
32 GLA Intelligence 2015: https://data.london.gov.uk/dataset/population-change-1939-2015 (impression: 2 February 2020).
33 Bloomberg, 26 January 2020: https://www.bloomberg.com/opinion/articles/2020-01-26/superstar-cities-london-new-york-amsterdam-are-losing-locals (impression: 14 February 2020).
34 Smith, P. D., *City: A Guidebook for the Urban Age*, London, Berlin, Sydney and New York, 2012, p. 312.
35 World Bank: https://data.worldbank.org/indicator/SP.RUR.TOTL.ZS?locations=ZG (impression: 2 February 2020).
36 World Population Review, 17 February 2020: https://worldpopulationreview.com/world-cities/lagos-population/ (impression: 22 March 2020).
37 McDougall, Robert and Kristiansen, Paul and Rader, Romina, 'Small scale agriculture results in high yields but requires judicious management of inputs to achieve sustainability', PNAS, 116 (1), 2019, pp. 129–34.

38 City Monitor, 18 June 2015: https://citymonitor.ai/government/granting-planning-permission-massively-increases-land-values-shouldnt-state-get-share-1154 (impression: 20 July 2021).

39 i24, 27 July 2019: https://www.i24news.tv/en/news/international/europe/1564227612-uk-s-johnson-vote-to-leave-eu-not-just-against-brussels-but-against-london-too (impression: 10 September 2020).

40 Davis, op. cit., p. 5.

41 Smith, op. cit., p. 312.

42 *Financial Times*, 23 March 2020: https://www.ft.com/content/1df725c0-6adb-11ea-800d-da70cff6e4d3 (impression: 23 March 2020).

43 Davis, op. cit., p. 5.

44 ONS: https://www.ons.gov.uk/peoplepopulationandcommunity/birthsdeathsandmarriages/lifeexpectancies/bulletins/lifeexpectancyatbirthandatage65bylocalareasinenglandandwales/2015-11-04#regional-life-expectancy-at-birth (impression: 14 February 2020).

45 TTN, 22 February 2017: https://timesofindia.indiatimes.com/city/kolkata/bengal-fertility-rate-lowest-in-country/articleshow/57283418.cms (impression: 21 August 2019).

## CHAPTER 4: FERTILITY

1 Five Stars and a Moon, 8 January 2016 http://www.fivestarsandamoon.com/2016/01/why-you-shouldnt-have-kids-in-singapore/ (impression: 29 January 2019).

2 Bricker, Darrell and Ibbitson, John, *Empty Planet: The Shock of Global Population Decline*, London, Robinson, 2019.

3 Levin, Hagai et al., 'Temporal Trends in Sperm Count: A Systematic Review and Meta-Regression Analysis', *Human Reproduction Update*, 23 (6), 2017, pp. 646–59.

4 NHS: https://www.nhs.uk/conditions/infertility/#:~:text=Infertility%20is%20when%20a%20couple,couples%20may%20have%20difficulty%20conceiving. (impression: 2 October 2020).

5 Government of Singapore https://www.singstat.gov.sg/modules/infographics/total-fertility-rate (impression: 22 October 2021); Mothership: https://mothership.sg/2018/04/singapore-total-fertility-rate-official/ (impression: 13 December 2018).

6 Morland, Paul, *The Human Tide: How Population Shaped the Modern World*, London, John Murray, 2019, pp. 90, 93.

7 Yap, Mui Teng, 'Fertility and Population Policy: the Singaporean Experience', *Journal of Population and Social Security Population*, (1) Suppl., 2003, p. 646.
8 Ibid., p. 651.
9 Ibid., p. 652.
10 *Straits Times*, 28 September 2018: https://www.straitstimes.com/singapore/spores-fertility-rate-down-as-number-of-singles-goes-up (impression: 29 March 2019).
11 *Straits Times*, 26 September 2016: https://www.straitstimes.com/singapore/fewer-sporean-babies-born-out-of-wedlock (impression: 29 March 2019), Yale Global Online: https://yaleglobal.yale.edu/content/out-wedlock-births-rise-worldwide (impression: 29 March 2019).
12 French, Marilyn, *The Women's Room*, London, André Deutsch, 1978, p. 47.
13 Bongaarts, John and Sobatka, Tomáš, 'A Demographic Explanation for the Recent Rise in European Fertility', *Population and Development Review*, 30 (1), 2012, pp. 83–120; The Austrian Academy of Sciences 2008: https://www.oeaw.ac.at/en/vid/data/demographic-data-sheets/european-demographic-data-sheet-2008/tempo-effect-and-adjusted-tfr/ (impression: 10 April 2019).
14 Morland, Paul, UnHerd, 17 October 2019: https://unherd.com/2019/10/has-hungary-conceived-a-baby-boom/ (impression: 16 October 2019).
15 Morland, Paul, *The Human Tide: How Population Shaped the Modern World*, London, John Murray, 2019, pp. 166–73.
16 Lieven, Dominic, *Towards the Flame: Empire, War and the End of Tsarist Russia*, London, Allen Lane, 2015, p. 60.
17 *Financial Times*, 12 March 2019: https://www.ft.com/content/f34bb0b0-2f8b-11e9-8744-e7016697f225 (impression: 3 April 2019).
18 UN Population Division: Data is for 2010–2015 period.
19 *Financial Times*, 24 August 2020: https://www.ft.com/content/c1bd20d6-f019-40ba-9ee7-b23e6150bf6c (impression: 24 August 2020).
20 Population Reference Bureau: https://interactives.prb.org/2021-wpds/asia/#east-asia (impression: 3 September 2021).
21 Statistics derived from the following sources: Fertility – US Government: https://www.cdc.gov/nchs/data/nvsr/nvsr68/nvsr68_01-508.pdf, Religiosity – Pew Research: https://www.pewresearch.org/fact-tank/2016/02/29/how-religious-is-your-state/?state=alabama, Voting – *New York Times*: https://www.nytimes.com/elections/2016/results/president, Income – Statista: https://www.statista.com/statistics/248063/per-capita-us-real-gross-domestic-product-gdp-by-state/ (all impressions: 5 April 2019).

22 Next Door Mom, 20 April 2011: http://www.nextdoormormon.com/2011/04/20/why-do-all-the-mormons-i-know-have-so-many-kids/ (impression: 5 April 2019).
23 Medium, 8 February 2018: https://medium.com/migration-issues/how-long-until-were-all-amish-268e3d0de87#:~:text., (impression: 2 October 2020).
24 Deseret News, 26 December 2019: https://www.deseret.com/indepth/2019/12/26/21020015/demographic-transition-fertility-rate-slowing-births-us-motherhood (impression: 21 August 2020).
25 Medium, op. cit.
26 Kaufmann, Eric, *Shall the Religious Inherit the Earth? Demography and Politics in the Twenty-First Century*, Profile Books, London, 2010, p. 35; Evans, Simon N. and Peller, Peter, 'A Brief History of Hutterite Demography', *Great Plains Quarterly*, 35, 1, 2015, pp. 79–101.
27 *Times of Israel*, 21 June 2018: https://www.timesofisrael.com/ultra-orthodox-reverse-uk-jewish-population-decline-study-finds/ (impression: 12 March 2019).
28 World Population Review https://worldpopulationreview.com/us-cities/kiryas-joel-ny-population (impression: 22 October 2021).
29 *Financial Times*, 7 April 2019: https://www.ft.com/content/dae642aa-5601-11e9-a3db-1fe89bedc16e (impression: 9 April 2019).
30 Schellekens, Jona and Anson, Jon, eds, *Israel's Destiny: Fertility and Mortality in a Divided Society*, New Brunswick and London, Transaction Publishers, 2007.
31 Mercatornet, 19 February 2019: https://mercatornet.com/israel-is-having-far-more-babies-than-any-other-developed-country/24064/ (impression: 21 August 2020), Smith, Tom, *Jewish Distinctiveness in America, a Statistical Portrait*, 2005, p. 73: https://www.jewishdatabank.org/databank/search-results/study/617 (impression: 31 January 2020).
32 Kaa, D. J. van de, *Europe's Second Demographic Transition*, Washington DC, Population Reference Bureau, 1987.
33 Kaufmann, op. cit., p. 130.
34 Ibid., passim.
35 *New York Jewish Week*, 17 August 2016: https://jewishweek.timesofisrael.com/orthodox-dropouts-still-tethered-to-faith/ (impression: 23 March 2020).
36 *Guardian*, 27 February 2019: https://www.theguardian.com/environment/shortcuts/2019/feb/27/is-alexandria-ocasio-cortez-right-to-ask-if-the-climate-means-we-should-have-fewer-children (impression: 6 March 2020).

37 *Forbes*, 7 April 2019, https://www.forbes.com/sites/ericmack/2019/04/07/a-quarter-of-japanese-adults-under-40-are-virgins-and-the-number-is-increasing/?sh=56099a6b7e4d (impression22 October 2022).
38 *Italy Magazine*, 12 April 2008: https://www.italymagazine.com/italy/science/less-sex-italian-couples-drop-male-sex-drive-blamed (impression: 9 April 2019).
39 *Time*, 26 October 2018: http://time.com/5297145/is-sex-dead/ (impression: 9 April 2019).
40 Kornich, Sabion, Brines, Julie and Leupp, Katrina, 'Egalitarianism, Housework and Sexual Frequency in Marriage', *American Sociological Review*, 78 (1), 2012, pp. 26–50.
41 Ibid.
42 Martine, George: 'Brazil's Fertility Decline 1965–1995: A Fresh Look at Key Factors', pp. 169–207 in Martine, George, Das Gupta, Monica and Chen, Lincoln C., eds, *Reproductive Change in India and Brazil*, Delhi and Oxford, Oxford University Press, 1998.
43 Birley, Daniel A., Tropf, Felix C. and Mills, Melinda C., 'What Explains the Heritability of Completed Fertility? Evidence from Two Large Twin Studies', *Behaviour Genetics*, 47, 2017, pp. 36–51; *Guardian*, 3 June 2015: https://www.theguardian.com/science/2015/jun/03/genetics-plays-role-in-deciding-at-what-age-women-have-first-child-says-study (impression: 2 October 2020).
44 Rosling, Hans, TED Talks: https://www.ted.com/talks/hans_rosling_religions_and_babies/transcript (impression: 21 December 2018).
45 United Nations Population Division 2017 Revisions (median fertility estimate).
46 Statistics Times, 12 September 2015: http://statisticstimes.com/economy/china-vs-india-gdp.php (impression: 10 April 2019).
47 Nippon.com, 25 July 2019, https://www.nippon.com/en/japan-data/h00438/japan-judged-low-on-happiness-despite-longevity.html (impression: 22 October 2021).
48 Bricker and Ibbitson, op. cit., passim.

## CHAPTER 5: AGEING

1 Statistical Institute of Catalonia: https://www.idescat.cat/pub/?id=aec&n=285&lang=en (impression: 27 October 2020). For an explanation of the median age, see Chapter 1 above.

2 *Guardian*, 14 November 2019: https://www.theguardian.com/world/2019/nov/14/second-death-in-hong-kong-protests-as-xi-demands-end-to-violence (impression: 24 August 2020).
3 BBC, 23 December 2017: https://www.bbc.com/news/world-asia-china-42465516 (impression: 25 August 2020).
4 Statistical Institute of Catalonia: https://www.idescat.cat/pub/?id=aec&n=285&lang=en (impression: 23 August 2019).
5 Statista: https://www.statista.com/statistics/275398/median-age-of-the-population-in-spain/ (impression: 23 August 2019). Note that this source gives 27.5 as the median age in Spain in 1950; given the trend, it is highly likely that it was below 25 in the 1930s.
6 According to the UN data, the only country where the median age was lower in 2020 than it had been in 2015 was Germany, presumably because of the large influx of young migrants in 2015 but not counted in that year's data. With its particularly heavy toll on the elderly, Covid-19 could have a similar albeit temporary effect on a much wider scale.
7 See for example Cincotta, Richard P., 'Demographic Security Come of Age', *ESCP Report*, 10, 2004, pp. 24–9; Urdal, Henrik R., 'A Clash of Generations? Youth Bulges and Political Violence', *International Studies Quarterly*, 50, 2006, pp. 607–29; Leuprecht, Christian, 'The Demography of Interethnic Violence', paper presented to the American Political Science Association, 2007. But see Guinnane, Timothy, 'The Human Tide: A Review Essay', *Journal of Economic Literature* 59 (4), 2021, p. 1330.
8 Leahy, Elizabeth et al., *The Shape of Things to Come: Why Age Structure Matters to a Safer, more Equitable World*, Washington DC, Population Action International, 2007.
9 Staveteig, Sarah, 'The Young and Restless: Population Age Structure and Civil War', in *Population and Conflict, Exploring the Links*, edited by Dalbeko, Geoffrey D. et al., Woodrow Wilson Center for Scholars, Environmental Change and Security Program report 11, 2005, pp. 12–19; Fearon, James D. and Laitin, David D., 'Sons of the Soil, Migrants and Civil War', *World Development*, 39 (2), 2010, pp. 199–211
10 Staveteig, op. cit.
11 Statista: https://www.statista.com/statistics/454349/population-by-age-group-germany/ (impression: 2 October 2020).
12 BBC, 10 August 2015: https://www.bbc.co.uk/news/newsbeat-33713015 (impression: 14 December 2020).

13 *Independent*, 18 February 2016: https://www.independent.co.uk/news/science/why-areteenagers-so-moody-a6874856.html (impression: 6 March 2020).
14 Johnson, Sara B., Blum, Robert W. and Giedd, Jay N., 'Adolescent Maturity and the Brain: The Promise and Pitfalls of Neuroscience Research in Adolescent Health Policy', *Journal of Adolescent Health*, 45 (3), 2009, pp. 216–21.
15 Brake: http://www.brake.org.uk/news/15-facts-a-resources/facts/488-young-drivers-the-hard-facts (impression: 8 March 2020).
16 Regev, Shirley, Rolison, Jonathan J. and Moutari, Salissou, 'Crash risk by driver age, gender, and time of day using a new exposure methodology', *Journal of Safety Research*, 66, 2018, pp. 131–40.
17 Mulderig, M. Chloe, 'An Uncertain Future: Youth Frustration and the Arab Spring', *Boston University Pardee Papers*, 16 2013, pp. 15, 23 and passim.
18 *Guardian*, 19 March 2014: https://www.theguardian.com/world/2014/mar/19/growing-youth-population-fuel-political-unrest-middle-east-south-america (impression: 8 March 2020).
19 *Jerusalem Post*, 4 September 2019: https://www.jpost.com/Opinion/Hezbollahs-demographic-problem-explains-its-restraint-600568 (impression: 8 March 2020).
20 See UN Population Division, https://population.un.org/wpp/Download/Standard/Population/ (impression: 24 October 2021).
21 *Pacific Standard*, 14 July 2017: https://psmag.com/social-justice/pax-americana-geriatrica-4416 (impression: 24 October 2021).
22 Morland, Paul, *Demographic Engineering: Population Strategies in Ethnic Conflict*, Farnham, Ashgate, 2014.
23 Ibid., passim.
24 Ceterchi, Ioan, Zlatescu, Victor, Copil, Dan, and Anca, Peter, *Law and Population Growth in Romania*, Bucharest, Legislative Council of the Socialist Republic of Romania, 1974.
25 Gatrell, Peter, *The Unsettling of Europe: The Great Migration, 1945 to the Present*, London, Allen Lane, 2019.
26 King, Leslie, 'Demographic Trends, Pro-Natalism and Nationalist Ideologies', *Ethnic and Racial Studies*, 25 (3), 2002, pp. 21–51.
27 Morland, Paul, *Demographic Engineering*, pp. 99–109.
28 Ibid., pp. 93–8.
29 *Guardian*, 7 March 2018: https://www.theguardian.com/media/2018/mar/07/nme-ceases-print-edition-weekly-music-magazine (impression: 25

March 2019); 2018 Cruise Industry Overview: https://www.f-cca.com/downloads/2018-Cruise-Industry-Overview-and-Statistics.pdf (impression: 25 March 2020).

30 BBC News, 15 March 2019: https://www.bbc.co.uk/news/uk-england-london-35126667 (impression: 13 March 2020).

31 *Guardian*, 25 June 2018: https://www.theguardian.com/film/2018/jun/25/hatton-garden-job-v-king-of-thieves-trailers-michael-caine (impression: 13 March 2020).

32 *Evening Standard*, 4 February 2021, https://www.standard.co.uk/news/crime/half-of-london-knife-crime-carried-out-by-teenagers-and-children-as-young-as-ten-police-figures-reveal-a4056596.html (impression: 24 October 2021).

33 World Atlas: https://www.worldatlas.com/articles/murder-rates-by-country.html (impression: 23 August 2019).

34 *Evening Standard*, 23 June 2018: https://www.standard.co.uk/news/crime/revealed-the-boroughs-with-the-highest-and-lowest-murder-rates-in-london-a3869671.html (impression: 29 January 2019); CBRE London Living 2016: https://www.cbreresidential.com/uk/sites/uk-residential/files/CBRE0352%20%20Borough%20by%20Borough%202016.pdf (impression: 29 January 2019).

35 E.g. Kahn, Samuel, 'Reconsidering the Donohue–Levitt Hypothesis', *American Catholic Philosophical Quarterly*, September 2016, pp. 583–620.

36 Griffith, Gwyn and Norris, Gareth, 'Explaining the Crime Drop: Contributions to Declining Crime Rates from Youth Cohorts since 2005', *Crime, Law and Social Change*, 73, 2020, pp. 25–53.

37 Dyson, Tim and Wilson, Ben, 'Democracy and the Demographic Transition', LSE Research Online, 2016: http://eprints.lse.ac.uk/66620/1/Wilson_Democracy%20and%20the%20demographic%20transition.pdf (impression: 25 September 2020).

## CHAPTER 6: OLD AGE

1 There is probably a slightly greater number of centenarians in the US, but it has a population two-and-half times that of Japan. China has a similar number, but in a population more than ten times larger.

2 *Washington Post*, 27 July: https://www.washingtonpost.com/news/worldviews/wp/2018/07/27/after-a-life-filled-with-sushi-and-calligraphy-worlds-oldest-person-dies-at-117/ (impression: 3 April 2020).

3 *Guinness Book of World Records*, 21 January 2019: https://www.guinnessworldrecords.com/news/2019/1/worlds-oldest-man-masazo-nonaka-dies-at-his-home-in-japan-aged-113-556396/ (impression: 27 August 2019).
4 *Jewish Chronicle*, 3 April 2020, p. 41.
5 *Prospect*, May 2020, p. 8.
6 Zak, Nikolay, Jeanne Calment: The Secret of Longevity: https://www.researchgate.net/publication/329773795_Jeanne_Calment_the_secret_of_longevity (impression: 7 April 2020).
7 National Geographic, 6 April 2017: https://www.nationalgeographic.com/books/features/5-blue-zones-where-the-worlds-healthiest-people-live/ (impression: 7 April 2020).
8 For a full explanation of life expectancy, see Morland, Paul, *The Human Tide: How Population Shaped the Modern World*, London, John Murray, 2020, pp. 283–5; for a somewhat simplified explanation, see the 2019 version of the same work.
9 Gratton, Lynda and Scott, Andrew, *The 100-Year Life: Living and Working in an Age of Longevity*, London, Bloomsbury Business, 2017, p. 26.
10 CNA, 24 April 20219: https://www.channelnewsasia.com/news/commentary/japan-ageing-population-old-harassing-young-working-age-11471252 (impression: 14 April 2020).
11 *Financial Times*, 23 April 2019: https://www.ft.com/content/b1369286-60f4-11e9-a27a-fdd51850994c (impression: 14 April 2020).
12 For an example of the growing literature linking stagnant economies with low fertility and slow population growth, see Jones, Charles I., *The End of Economic Growth? Unintended Consequences of a Declining Population*, Stanford NBER, 2020.
13 World Economic Forum: https://www.weforum.org/agenda/2019/02/japan-s-workforce-will-shrink-20-by-2040/ (impression: 31 March 2020).
14 Macrotrends: https://www.macrotrends.net/2593/nikkei-225-index-historical-chart-data (impression: 31 March 2020).
15 World Bank: https://data.worldbank.org/indicator/NY.GDP.MKTP.KD.ZG?locations=JP (impression: 31 March 2020).
16 Macrotrends: https://www.macrotrends.net/countries/JPN/japan/inflation-rate-cpi (impression: 31 March 2020).
17 ONS https://www.ons.gov.uk/employmentandlabourmarket/peopleinwork/employmentandemployeetypes/timeseries/bbfw/lms (impression: 24 October 2021).
18 Vollrath, Dietrich, *Fully Grown: Why a Stagnant Economy is a Sign of Success*, Chicago and London, University of Chicago Press, 2020, p. 63.

19 *Financial Times*, 17 October 2020: https://www.ft.com/content/8b2fbf82-8cbe-487e-af63-b3b006f9672d (impression: 18 October 2020).

20 See for example Kelton, Stephanie, *The Deficit Myth: Modern Monetary Theory and the Birth of the People's Economy*, New York, Public Affairs, 2020. Kelton's work provides a thorough outline of the theory and a justification for it but does not make the argument that it is required now in a way it has not been in the past because of demography.

21 For share of wealth by cohort see *Washington Post*, 13 December 2019: https://www.washingtonpost.com/business/2019/12/03/precariousness-modern-young-adulthood-one-chart/ (impression: 14 December 2020).

22 Goodhart, Charles and Pradhan, Manoj, *The Great Demographic Reversal: Ageing Societies, Waning Inequality and an Inflation Revival*, Cham Switzerland, Palgrave Macmillan, 2020.

23 A Measured View of Healthcare: https://measuredview.wordpress.com/2014/10/07/15/ (impression: 3 April 2020).

24 Global Spending on Health: A World Transition, World Health Organization 2019, p. 6: https://www.who.int/health_financing/documents/health-expenditure-report-2019.pdf?ua=1 (impression: 14 April 2019).

25 Forbes, 6 March 2020: https://www.forbes.com/sites/stephenpope/2020/03/06/migrating-european-youth-threatens-europes-pension-program/ (impression: 3 April 2020).

26 *The Gerontologist*, 54 (1), February 2014: https://academic.oup.com/gerontologist/article/54/1/5/561938 (impression: 3 April 2020).

27 As to whether pay-as-you-go welfare is a Ponzi scheme, it may appear so on the face of the matter. But once again, it is worth referring to the proponents of Modern Monetary Theory who argue that, providing the country is able to produce the resources and services required without inflation or unfinanceable trade deficits, it is not. This is a debate I am happy to leave to the economist.

28 *New York Times*, 11 January 2020: https://www.nytimes.com/2020/01/11/world/europe/france-pension-protests.html (impression: 3 April 2020).

29 Goodhart and Pradhan, op. cit., pp. 49–50.

30 Eurostat: https://ec.europa.eu/eurostat/statistics-explained/index.php?title=Ageing_Europe_-_statistics_on_working_and_moving_into_retirement (impression: 24 October 2021).

31 Rest Less / ONS 27 May 2019: https://restless.co.uk/press/the-number-of-over-70s-still-working-has-more-than-doubled-in-a-decade/ (impression: 12 April 2020).

32 *Washington Post*, 30 March 2020: https://www.washingtonpost.com/business/2020/03/30/retail-workers-their-60s-70s-80s-say-theyre-worried-about-their-health-need-money/ (impression: 1 September 2020).

33 British Election Study, 12 February 2018: https://www.britishelectionstudy.com/bes-impact/youthquake-a-reply-to-our-critics/#.XpLlI25FyUm (impression: 12 April 2020).

34 *Nature*, 28 August 2020: https://www.nature.com/articles/d41586-020-02483-2 (impression: 19 October 2020).

35 *Guardian*, 7 October 2019: https://www.theguardian.com/commentisfree/2019/oct/27/age-rather-than-class-now-determines-how-britain-votes (impression: 5 April 2020).

36 Lord Ashcroft Polls, 15 March 2019: https://lordashcroftpolls.com/2019/03/a-reminder-of-how-britain-voted-in-the-eu-referendum-and-why/ (impression: 12 April 2020).

37 For the American case see *Washington Post*, 11 February 2019: https://www.washingtonpost.com/news/monkey-cage/wp/2019/02/11/yes-young-people-voted-at-higher-rates-in-2018-but-so-did-every-age-group/ (impression: 25 September 2020).

38 Pew Research Center, 9 August 2018: https://www.people-press.org/2018/08/09/an-examination-of-the-2016-electorate-based-on-validated-voters/ (impression: 5 April 2020).

39 *Guardian*, 5 November 2020: https://www.theguardian.com/us-news/2020/nov/05/us-election-demographics-race-gender-age-biden-trump (impression: 14 December 2020).

40 *Independent*, 7 March 2017: https://www.independent.co.uk/news/nearly-half-young-french-voters-marine-le-pen-emmanuel-macron-french-election-2017-a7723291.html (impression: 12 April 2020).

41 This Retirement Life, 20 February 2020: https://thisretirementlife.com/2020/02/28/retiring-to-costa-rica/ (impression: 16 October 2020).

42 *The Economist*, 4 April 2020, p. 45.

43 *Financial Times*, 13 December 2019: https://www.ft.com/content/b909e162-11f6-44f3-8eab-ebc48d8c6976 (impression: 13 April 2020).

44 UN Desa, 25 February 2019: https://www.un.org/en/development/desa/population/events/pdf/expert/29/session3/EGM_25Feb2019_S3_VipanPrachuabmoh.pdf (impression: 13 April 2020).

45 *Bangkok Post*, 11 December 2018: https://www.bangkokpost.com/life/social-and-lifestyle/1591554/how-the-old-stay-young (impression: 11 December 2018).

46 *Wall Street Journal*, p. B4, 14 January 2019: https://assets.website-files.com/5b036b7ed0a90fe56e35e376/5c771db8776d024333636dcc_elder-care-in-japan-propels-innovation.pdf (impression: 15 April 2020).
47 *Independent*, 9 April 2019: https://www.independent.co.uk/arts-entertainment/photography/japan-robot-elderly-care-ageing-population-exercises-movement-a8295706.html (impression: 14 April 2020).
48 Ibid.
49 Health Equity in England: The Marmot Review 10 Years On: Institute of Health Equity, pp. 15–18: https://www.health.org.uk/sites/default/files/upload/publications/2020/Health%20Equity%20in%20England_The%20Marmot%20Review%2010%20Years%20On_full%20report.pdf (impression: 14 April 2020).
50 ONS https://www.ons.gov.uk/peoplepopulationandcommunity/birthsdeathsandmarriages/lifeexpectancies/articles/ethnicdifferencesinlifeexpectancyandmortalityfromselectedcausesinenglandandwales/2011to2014 (impression: 24 October 2021).
51 Continuous Mortality Investigation Briefing Note 2018: https://www.actuaries.org.uk/system/files/field/document/CMI%20WP119%20v01%202019-03-07%20-%20CMI%20Mortality%20Projections%20Model%20CMI_2018%20Briefing%20Note.pdf; *Financial Times*, 1 March 2018: https://www.ft.com/content/dc7337a4-1c91-11e8-aaca-4574d7dabfb6 (impressions: 14 April 2020).
52 Cavendish, Camilla, *Extra Time: 10 Lessons for an Ageing World*, London, HarperCollins, 2019, p. 24.
53 Dorling, Danny and Gietel-Basten, Stuart, *Why Demography Matters*, Cambridge, Polity, 2018, p. 49.

## CHAPTER 7: POPULATION DECLINE

1 Making the History of 1989, item #310: http://chnm.gmu.edu/1989/items/show/319 (impression: 17 December 2018).
2 Öktem, Kerem, 'The Nation's Imprint: Demographic Engineering and the Change of Toponymes in Republican Turkey', *European Journal of Turkish Studies*, 7, 2008, passim.
3 Morland, Paul, *The Human Tide: How Population Shaped the Modern World*, London, John Murray, 2019, p. 188.
4 Note that Bulgarian women are in fact having their children relatively early, with the average first birth at around twenty-six.

5 Euractiv, 26 December 2019: https://www.euractiv.com/section/economy-jobs/news/alarming-low-birth-rates-shut-down-schools-in-greece/ (impression: 17 April 2020).
6 *Financial Times*, 15 October 2020: https://www.ft.com/content/5dafc7e1-d233-48c4-bd6b-90a2ed45a6e7 (impression: 16 October 2020).
7 DW, 25 November 2018: https://www.dw.com/en/germanys-lonely-dead/a-46429694 (impression: 27 August 2019).
8 Politico, 6 January 2016, https://www.politico.eu/article/germany-set-immigration-record-in-2015/ (impression: 25 October 2021).
9 *Guardian*, 7 May 2019: https://www.theguardian.com/cities/2019/may/07/reversing-the-brain-drain-how-plovdiv-lures-young-bulgarians-home (impression: 17 April 2020).
10 Caritas Bulgaria, *The Bulgarian Migration Paradox: Migration and Development in Bulgaria*, 2019: https://www.caritas.eu/wordpress/wp-content/uploads/2019/06/CommonHomeBulgariaEN.pdf p.7 (impression: 17 April 2019).
11 *Guardian*, 7 May 2019, op. cit.
12 *Financial Times*, 15 October 2020, op. cit.
13 Keen, M. H, *England in the Later Middle Ages: A Political History*, Routledge, London and New York, 1973, p. 170.
14 Outram, Quentin, 'The Demographic Impact of Early Modern Warfare', *Social Science History*, 26 (2), 2002, pp. 245–72, 248.
15 Lee, Harry F. and Zhang, David D., 'A Tale of Two Population Crises in Recent Chinese History', *Climatic Change*, 2013, 116, pp. 285–308; Liebmann, Matthew J., Farella, Joshua, Roos, Christopher I., Stack, Adam, Martini, Sarah and Swetnam, Thomas W., 'Native American Depopulation, Forestation and Fire Regimes in South West United States 1492–1900', *PNAS*, 113 (6), 2013, pp. 696–704.
16 Georgieva-Stankova, N., Yarkova, Y. and Mutafov, E., 'Can Depopulated Villages Benefit from the Social and Economic Incorporation of Ethnic and Immigrant Communities? A Survey from Bulgaria', *Trakia Journal of Sciences*, 16 (2), 2018, p. 140.
17 Mladenov, Čavdar and Ilieva, Margarita, 'The Depopulation of the Bulgarian Villages', *Bulletin of Geography: Socio Economic Series*, 17, 2012, p. 100.
18 BBC News, 17 September 2017: https://www.bbc.co.uk/news/world-europe-41109572 (impression: 27 August 2019).
19 Balkan Insight, 26 February 2020: https://balkaninsight.com/2020/02/26/where-did-everyone-go-the-sad-slow-emptying-of-bulgarias-vidin/ (impression: 27 April 2020).

20 NBC News, 14 May 2019, https://www.nbcnews.com/news/world/russia-s-dying-villages-inspire-rising-star-art-world-n994436 (impression: 24 October 2021).
21 Radio Free Europe / Radio Liberty, 10 December 2018: https://www.rferl.org/a/russia-shelepovo-dying- village/29648412.html (impression: 20 April 2020).
22 Russia Matters, 13 September 2019: https://www.russiamatters.org/blog/russian-population-decline-spotlight-again (impression: 20 April 2020).
23 Interview with Emily Ferris, Research Fellow, Royal United Services Institute, 21 April 2020; SCMP: https://www.scmp.com/week-asia/geopolitics/article/2100228/chinese-russian-far-east-geopolitical-time-bomb (impression: 21 April 2020).
24 *South China Morning Post*, 18 April 2018: https://www.scmp.com/news/china/society/article/2142363/rural-exodus-leaves-shrinking-chinese-village-full-ageing-poor (impression: 21 April 2020).
25 National Bureau of Statistics of China: http://www.stats.gov.cn/english/PressRelease/202105/t20210510_1817185.html (impression: 9 September 2021).
26 *The Economist*, 1 May 2021, pp. 48–9.
27 *Guardian*, 13 June 2016: https://www.theguardian.com/world/2016/jun/13/warning-four-killed-bear-attacks-akita-japan (impression: 17 December 2018).
28 BBC, 31 October 2019: https://www.bbc.com/worklife/article/20191023-what-will-japan-do-with-all-of-its-empty-ghost-homes (impression: 21 April 2020).
29 Brickunderground, 24 August 2015: https://www.brickunderground.com/blog/2015/08/japanese_suburbs_are_the_polar_opposites_of_ (impression: 24 October 2021).
30 A Vision of Britain Through Time: https://www.visionofbritain.org.uk/unit/10217647/cube/AGESEX_85UP (impression: 22 April 2020).
31 Stoke on Trent Live, 16 January 2020: https://www.stokesentinel.co.uk/news/stoke-on-trent-news/stoke-trent-pubs-decline-numbers-3744849 (impression: 22 April 2020).
32 A Vision of Britain Through Time: https://www.visionofbritain.org.uk/unit/10217647/cube/TOT_POP (impression: 8 October 2020).
33 Lane, Laura, Grubb, Ben and Power, Anne, 'Sheffield City Story', *LSE Centre for Analysis of Social Exclusion*, 2016, pp. 4, 14.
34 *Financial Times*, 25 August 2019: https://www.ft.com/content/c88b4c54-b925-11e9-96bd-8e884d3ea203 (impression: 22 April 2020).

35 Bricker, Darrell and Ibbitson, John, *Empty Planet: The Shock of Global Population Decline*, New York, Crown, 2019, p. 172.
36 Morland, Paul, *The Human Tide: How Population Shaped the Modern World*, London, John Murray, p. 89; *Pittsburgh Post-Gazette*, 24 March 2020:https://www.post-gazette.com/opinion/Op-Ed/2019/03/24/The-eternal-fear-of-race-suicide/stories/201903240066 (impression: 26 April 2020).
37 Sabin, Paul, *The Bet: Paul Ehrlich, Julian Simon, and Our Gamble over the Earth's Future*, New Haven and London, Yale University Press, 2013, p. 22.
38 NBS (Nigeria), 2017 Demographic Statistics Bulletin, May 2018, p. 10.
39 Bricker and Ibbitson, op. cit., p. 68.
40 See for example Webb, Stephen, *If the Universe is Teeming with Aliens, Where is Everybody? Fifty Solutions to the Fermi Paradox and the Problem of Extraterrestrial Life*, Copernicus Books, New York, 2002.
41 *Financial Times*, 9 June 2019: https://www.ft.com/content/05baa6ae-86dd-11e9-a028-86cea8523dc2 (impression: 2 September 2019).

## CHAPTER 8: ETHNIC CHANGE

1 Kidsdata: https://www.kidsdata.org/topic/36/school-enrollment-race/table#fmt (impression: 2 September 2021).
2 Lewis, Edward R., *America – Nation or Confusion? A Study of our Immigration Problems*, New York and London, Harper and Brothers, 1928, p. 13.
3 Kaufmann, Eric, *The Rise and Fall of Anglo-America*, Cambridge, Mass., Harvard University Press, 2004.
4 Morland, Paul, *Demographic Engineering: Population Strategies in Ethnic Conflict*, Farnham, Ashgate, 2014, pp. 149–51.
5 Lepore, Jill, *These Truths: A History of the United States*, London and New York, W.W. Norton, 2018, p. 468.
6 Public Policy Institute for California: https://www.ppic.org/publication/californias-population/ (impression: 27 August 2019).
7 Public Policy Institute of California: https://www.ppic.org/publication/californias-population/ (impression: 1 May 2020).
8 Kidsdata, op. cit.
9 US Census: https://www.census.gov/quickfacts/TX (impression: 8 September 2020).
10 Texas Demographic Center 14 September 2017: https://demographics.texas.gov/Resources/Presentations/OSD/2017/2017_09_14_DepartmentofSavingsandMortgageLending.pdf (impression: 8 September 2020).

11 Brookings Institute, 14 March 2018: https://www.brookings.edu/blog/the-avenue/2018/03/14/the-us-will-become-minority-white-in-2045-census-projects/ (impression: 1 May 2020).
12 *New York Times*, 8 June 2019: https://www.nytimes.com/2019/06/08/us/politics/migrants-drown-rio-grande.html (impression: 4 May 2020).
13 CBS, 26 June 2019: https://www.cbsnews.com/news/tragic-photo-migrant-father-oscar-alberto-martinez-ramirez-toddler-who-died-trying-to-cross-the-rio-grande/ (impression: 4 May 2019).
14 Eschbach, Karl, Hagan, Jacqueline, Rodriguez, Nestor, Hérnandez-Léon, Rubén and Bailey, Stanley, 'Death at the Border', *International Migration Review*, 33 (2), 1999, pp. 430–54.
15 Darwin, Charles, *The Descent of Man and Selection in Relation to Sex*, New York, Appleton and Company, 1871, p. 193.
16 *Guardian*, 2 September 2015: https://www.theguardian.com/world/2015/sep/02/shocking-image-of-drowned-syrian-boy-shows-tragic-plight-of-refugees (impression: 5 May 2020).
17 *Sunday Times*, 22 August 2021, p. 25.
18 Pew Research, 2 August 2016: https://www.pewresearch.org/global/2016/08/02/number-of-refugees-to-europe-surges-to-record-1-3-million-in-2015/pgm_2016-08-02_europe-asylum-01/ (impression: 5 May 2020).
19 ONS: https://www.ons.gov.uk/peoplepopulationandcommunity/populationandmigration/internationalmigration/bulletins/ukpopulationbycountryofbirthandnationality/2017 (impression: 6 May 2020).
20 *The Times*, 9 May 2019: https://www.thetimes.co.uk/article/up-to-75-of-babies-are-born-to-migrant-mothers-in-parts-of-uk-j2xv9r858 (impression: 5 May 2020).
21 2011 Census: A Profile of Brent: https://www.whatdotheyknow.com/request/520769/response/1251473/attach/11/Equalities%20Assement%20Document%208.pdf?cookie_passthrough=1 (impression: 25 September 2020).
22 DW.com, 4 October 2016: https://www.dw.com/en/record-rise-in-babies-with-foreign-mothers-in-germany/a-35952212 (impression: 5 May 2020).
23 Coleman, David, 'Projections of Ethnic Minority Population in the United Kingdom 2006–2056', *Population and Development Review*, 36 (3) 2010, pp. 456, 462.
24 Pew Research Center, 29 November 2017: http://www.pewforum.org/2017/11/29/europes-growing-muslim-population/ (impression: 17 December 2018).

25 Les Observateurs.ch, 28 September 2015: https://lesobservateurs.ch/2015/09/28/charles-de-gaulle-colombey-les-deux-mosquees/ (impression: 17 December 2018).
26 *New York Times*, 7 March 2019: https://www.nytimes.com/2019/03/07/us/us-birthrate-hispanics-latinos.html (impression: 3 May 2020).
27 Dubuc, Sylvie, 'Immigration to the UK from High Fertility Countries: Intergenerational Adaptation and Fertility Convergence', *Population and Development Review*, 38 (2), p. 358.
28 The Migration Observatory, 20 January2020: https://migrationobservatory.ox.ac.uk/resources/briefings/uk-public-opinion-toward-immigration-overall-attitudes-and-level-of-concern/ (impression: 7 May 2020).
29 British Social Attitudes: https://www.bsa.natcen.ac.uk/latest-report/british-social-attitudes-31/immigration/introduction.aspx (impression: 25 September 2020).
30 BBC, 28 April 2015: https://www.bbc.co.uk/news/election-2015-32490861 (impression: 7 May 2020).
31 Kaufmann, Eric, *Whiteshift: Populism, Immigration and the Future of White Majorities*, London, Allen Lane, 2018, pp. 201–4.
32 Dorling, Danny, *Slowdown: The End of the Great Acceleration and Why It's Good for the Economy, the Planet and Our Lives*, New Haven and London, Yale, 2020, pp. 153–4.
33 *Guardian*, 24 May 2019: https://www.theguardian.com/uk-news/2019/may/24/uk-government-misses-net-migration-target-for-37th-time-in-a-row (impression: 25 September 2020).
34 ONS: https://www.ons.gov.uk/peoplepopulationandcommunity/populationandmigration/internationalmigration/bulletins/migrationstatisticsquarterlyreport/november2019 (impression: 25 September 2020).
35 France 24, 21 April 2017: https://www.france24.com/en/20170420-france-presidential-history-looking-back-jean-marie-le-pen-thunderclap-election-shocker (impression: 7 September 2020); France 24: https://graphics.france24.com/results-second-round-french-presidential-election-2017/ (impression: 7 September 2020).
36 Morland, op. cit., pp. 53–83.
37 Ibid., p. 57.
38 Thatcher, Margaret, *The Downing Street Years*, London, Harper Collins, 1993, p. 385; Irish Central, 30 June 2013: https://www.irishcentral.com/news/margaret-thatcher-admitted-to-irish-roots-a-great-great-irish-

grandmother-at-1982-dinne-213737941-237760641 (impression: 15 May 2020).

39 Pew Research Center – Hispanic Trends, 20 December 2017: https://www.pewresearch.org/fact-tank/2019/08/08/hispanic-women-no-longer-account-for-the-majority-of-immigrant-births-in-the-u-s/ (impression: 8 September 2020).

40 Pew Research Center – Religion and Public Life, 17 October 2017: https://www.pewforum.org/2019/10/17/in-u-s-decline-of-christianity-continues-at-rapid-pace/ (impression: 8 September 2020).

41 Pew Research Center – Hispanic Identify Fades Across Generations as Immigrant Connections Fall Away, 20 December 2017: https://www.pewresearch.org/hispanic/2017/12/20/hispanic-identity-fades-across-generations-as-immigrant-connections-fall-away/ (impression: 14 June 2021).

## CHAPTER 9: EDUCATION

1 CountryEconomy.com: https://countryeconomy.com/demography/literacy-rate/bangladesh (impression: 23 August 2019).

2 Smil, Vaclav, *Growth: From Microorganisms to Megacities*, Cambridge, Mass., Massachusetts Institute of Technology, 2019, p. 429.

3 Our World Data: https://ourworldindata.org/how-is-literacy-measured (impression: 16 July 2020).

4 Ranjan, Amit, 'Bangladesh Liberation War of 1971: Narratives, Impacts and Actors', *Indian Quarterly*, 72 (2), 2016, p. 135; as Ranjan points out, there are divergent views on this number, with many believing the actual figure to be substantially lower.

5 The Forum: https://archive.thedailystar.net/forum/2008/march/basket.htm (impression: 14 July 2020); in fact it appears that it was not Kissinger but a more junior US official who used the term.

6 Banglapedia: http://en.banglapedia.org/index.php?title=Literacy (impression: 27 October 2019).

7 Bangladesh Bureau of Statistics http://bbs.portal.gov.bd/sites/default/files/files/bbs.portal.gov.bd/page/4c7eb0f0_e780_4686_b546_b4fa0a8889a5/BDcountry%20project_final%20draft_010317.pdf (impression: 23 August 2020).

8 Our World Data, 8 June 2018: https://ourworldindata.org/how-is-literacy-measured (impression: 16 July 2020).

9 World Concern, 19 December 2017: https://humanitarian.worldconcern.org/2017/12/19/girls-education-bangladesh/ (impression: 22 May 2020).
10 The Diplomat, December 2017: https://thediplomat.com/2017/12/bangladesh-empowers-women/ (impression: 22 May 2020)
11 Smil, op. cit., p. 305.
12 World Bank: https://data.worldbank.org/indicator/SE.TER.ENRR?locations=KR (impression: 27 October 2020).
13 Fact Maps: https://factsmaps.com/pisa-2018-worldwide-ranking-average-score-of-mathematics-science-reading/ (impression: 14 July 2020).
14 World Bank: https://data.worldbank.org/indicator/NY.GDP.MKTP.CD?most_recent_value_desc=true (impression: 14 July 2020).
15 Wolla, A. Scott and Sullivan, Jessica, 'Education, Income and Wealth' https://research.stlouisfed.org/publications/page1-econ/2017/01/03/education-income-and-wealth/ (impression: 25 October 2021).
16 Schwab, Klaus, and Sala i Martín, Xavier, *The Global Competitiveness Report 2017–2018*, World Economic Forum, 2017, p. 110.
17 Loveluck, Louisa, *Education in Egypt: Key Challenges*, Chatham House, March 2012.
18 Ghafar, Adel Abdel, *Educated but Unemployed: The Challenge Facing Egypt's Youth*, Washington and Doha, Brookings, 2016.
19 *The Economist*, 18 July 2020, p. 37.
20 Turchin, Peter, 'Political instability may be a contributor in the coming decade', *Nature*, 463, 2010, p. 608; *The Economist*, 24 October 2020, p. 76.
21 *Daily Star*, 11 October 2019: https://www.thedailystar.net/backpage/world-bank-latest-report-one-in-three-graduates-unemployed-in-bangladesh-1812070 (impression: 19 October 2020).
22 *Independent*, 22 November 2015: https://www.independent.co.uk/news/education/education-news/the-19-countries-with-the-highest-ratio-of-women-to-men-in-higher-education-a6743976.html (impression: 20 December 2020).
23 Kharas, Homi and Zhang, Christine, *Women in Development*, 21 March 2014, Brookings Institute: https://www.brookings.edu/blog/education-plus-development/2014/03/21/women-in-development/ (impression: 25 May 2020); Ugbomeh, George M. M., 'Empowering Women in Agricultural Education for Sustainable Rural Development', *Community Development Journal*, 36 (4), 2001, pp. 289–302.

24 Cornell Alliance for Science, December 2019: https://allianceforscience.cornell.edu/blog/2019/12/new-initiative-aims-to-empower-africas-female-farmers/ (impression: 15 July 2020).
25 Reimers, Malte and Klasen, Stephan, 'Revisiting the Role of Education for Agricultural Productivity', *American Journal of Agricultural Economics*, 95 (1), pp. 131–52, 2013.
26 Government of India – Ministry of Statistics and Programme Implementation, Literacy and Education http://www.mospi.gov.in/sites/default/files/reports_and_publication/statistical_publication/social_statistics/Chapter_3.pdf, p. 4 (impression: 15 July 2020).
27 Glaeser, Edward L., Ponzetto, Giacomo and Shleifer, Andrei, *Why Does Democracy Need Education?*, NBER Working Paper 12128: https://www.nber.org/papers/w12128.pdf (impression: 15 July 2020); Acemoglu, Daron, Johnson, Simon, Robinson, James A. and Yared, Pierre 2005, 'From Education to Democracy', *AEA Papers and Proceedings,* 95 (2): https://pubs.aeaweb.org/doi/pdf/10.1257/000282805774669916 (impression: 15 July 2020).
28 Case, Anne and Deaton, Angus, *Deaths of Despair and the Future of Capitalism*, Princeton and Oxford, Princeton University Press, 2020, pp. 57, 59, 66.
29 Harber, Clive, *Education and International Development: Theory, Practice and Issues*, Oxford, Symposium Books, 2014, p. 31.
30 Global Citizen, 18 June 2017: https://www.globalcitizen.org/en/content/rihanna-learned-challenges-facing-students-in-mala/ (impression: 19 October 2020).
31 Harber, op. cit., p. 72.
32 Allais, Stephanie Matseleng, 'Livelihood and Skills', in McCowan, Tristan and Unterhalter, Elaine, eds, *Education and International Development: An Introduction*, London, Bloomsbury, 2015, p. 248.
33 UPI, 26 August 2015: https://www.upi.com/Top_News/World-News/2015/08/26/86-percent-of-South-Korean-students-suffer-from-schoolwork-stress/8191440611783/#:~:text=The%20study%20habits%20among%20South,early%2C%20according%20to%20the%20survey.&text=But%20the%20system%20is%20taking,if%20they%20take%20a%20break. (impression: 15 July 2020).
34 Berkeley Political Review, 31 October 2017: https://bpr.berkeley.edu/2017/10/31/the-scourge-of-south-korea-stress-and-suicide-in-korean-society/ (impression: 15 July 2020).
35 Unterhalter, Elaine, 'Education and International Development: A History of the Field', in McCowan and Unterhalter, op. cit., p. 17.

36 Garnett Russell, Susan and Bajaj, Monisha, 'Schools, Citizens and Nation State', in McCowan and Unterhalter, op. cit., p. 103.
37 See for example Gellner, Ernest, *Nations and Nationalism*, Ithaca and New York, Cornell University Press, 1983.
38 Goodhart, David, *Hand, Head, Heart: The Struggle for Dignity and Status in the 21st-Century*, London, Penguin, 2020; Vollrath, Dietrich, *Fully Grown: Why a Stagnant Economy is a Sign of Success*, Chicago and London, University of Chicago Press, 2020, pp. 26–34.
39 HESA, 22 October 2019: https://www.hesa.ac.uk/news/22-10-2019/return-to-degree-research (impression: 8 October 2020).
40 UNESCO: http://uis.unesco.org/country/TD (impression: 14 July 2020).
41 UNESCO Fact Sheet 45, Literacy Rates Continue to Rise from One Generation to the Next, September 2017, pp. 7, 9.
42 UNESCO: http://uis.unesco.org/en/country/gq (impression: 15 July 2020).
43 *Financial Times*, 14 June 2018: https://www.ft.com/content/d110fbba-8b69–11e9-a1c1–51bf8f989972 (impression: 25 May 2020).
44 The work in question is Riley, Matthew and Smith, Anthony D., *Nation and Classical Music*, Woodbridge, The Boydell Press, 2016.

## CHAPTER 10: FOOD

1 World Bank, data for 1993 to 2018: https://data.worldbank.org/indicator/AG.PRD.CREL.MT?locations=ET (impression: 26 October 2021).
2 This further assumes unchanging mortality rates and age structures. Even if these further assumptions did not hold, the impact on the rate of population growth would not be material or affect the argument. The start of the Common Era is arbitrary; in principle, an earlier or later date for the exponential growth of humans could have been chosen.
3 Buck, Pearl S., *The Good Earth*, New York, Washington Square Press, 2005, p. 37.
4 *Huffington Post*, 6 September 2017: https://www.huffingtonpost.ca/development-unplugged/how-women-in-ethiopia-empower-communities-through-nutrition_a_23197349/ (impression: 7 June 2018).
5 Kidane, Asmeron, 'Mortality Estimates of the 1984–85 Ethiopian Famine', *Scandinavian Journal of Social Medicine*, 18 (4), 1990, pp. 281–6.

6 *Guardian*, 22 October 2014: https://www.theguardian.com/world/2014/oct/22/-sp-ethiopia-30-years-famine-human-rights (impression: 22 September 2020).
7 Knoema: https://knoema.com/atlas/Ethiopia/topics/Education/Literacy/Adult-literacy-rate (impression: 5 February 2019).
8 World Bank: https://data.worldbank.org/indicator/ag.yld.crel.kg (impression: 27 August 2019).
9 Global Nutrition Report: https://globalnutritionreport.org/resources/nutrition-profiles/africa/eastern-africa/ethiopia/ (impression: 16 September 2020).
10 Ibid.
11 Bourne, Joel K. Jr., *The End of Plenty: The Race to Feed a Crowded World*, Melbourne and London, Scribe, 2015, p. 79.
12 Earth Policy Institute, January 2013, http://www.earth-policy.org/indicators/C54 (impression: 16 September 2020); World Bank: https://data.worldbank.org/indicator/AG.PRD.CREL.MT (impression: 26 October 2021).
13 Anadolu Agency, 21 March 2016: https://www.aa.com.tr/en/todays-headlines/ethiopia-struggling-to-cope-with-deforestation/541174 (impression: 17 September 2020).
14 BBC, 11 August 2019: https://www.bbc.co.uk/news/world-africa-49266983 (impression: 17 September 2020).
15 Lindstrom, David P. and Woubalem, Zewdu, 'The Demographic Components of Fertility Decline in Addis Ababa, Ethiopia: A Decomposition Analysis', *Genus*, 59 (3/4), 2–3, 2003, p. 149.
16 UNEP: https://www.unenvironment.org/news-and-stories/story/towards-sustainable-desalination (impression: 1 October 2020); Advisian: https://www.advisian.com/en-gb/global-perspectives/the-cost-of-desalination# (impression: 1 October 2020).
17 Kumar, Amit et al., 'Direct Electrosynthesis of Sodium Hydroxide and Hydrochloric Acid from Brine Streams', *Nature Catalysis*, (2), 2019, pp. 106–13.
18 *Nature*, 28 July 2010: https://www.nature.com/articles/466531a (impression: 21 September 2020).
19 Woodruff, William, *America's Impact on the World: A Study of the Role of the United States in the World Economy, 1750–1970*, London, Macmillan, 1975, p. 38.
20 Collingham, Lizzie, *The Hungry Empire: How Britain's Quest for Food Shaped the Modern World*, London, The Bodley Head, 2017, pp. 220, 222.

21 For a full discussion, see Morland, Paul, *The Human Tide: How Population Shaped the Modern World*, London, John Murray, 2019, pp. 69–99.
22 Otter, Chris, *Diet for a Large Planet: Industrial Britain, Food Systems and World Ecology* (Chicago and London, Chicago University Press, 2020), pp. 48, 50.
23 New World Encyclopaedia: https://www.newworldencyclopedia.org/entry/War_of_the_Pacific (impression: 21 September 2020).
24 Charles, Daniel, *Between Genius and Genocide: The Tragedy of Fritz Haber, Father of Chemical Warfare*, London, Jonathan Cape, 2005, p. 73.
25 Smil, Vaclav, 'Detonator of the Population Explosion', *Nature*, 400, 29 July 1999, p. 415.
26 Smil, Vaclav, *Growth: From Microorganisms to Megacities*, Cambridge Mass., The MIT Press, 2019, p. 390.
27 Snyder, Timothy, *Black Earth: The Holocaust as History and Warning*, London, The Bodley Head, 2015, p. 10.
28 For a fuller discussion of this topic see Staudenmaier, Peter, 'Organic Farming in Nazi Germany: The Politics of Biodynamic Agriculture 1933–1945', *Environmental History*, 18 (2), 2013, pp. 383–411.
29 Bourne, op. cit., p. 74.
30 *Guardian*, 1 April 2014: https://www.theguardian.com/global-development/poverty-matters/2014/apr/01/norman-borlaug-humanitarian-hero-menace-society (impression: 16 September 2020).
31 Mackinac Center, 15 December 2009: https://www.mackinac.org/11516#:~:text=Gregg%20Easterbrook%20quotes%20Borlaug%20saying,suites%20in%20Washington%20or%20Brussels. (impression: 6 October 2020).
32 Sinha, Manish, 'The Bengal Famine of 1943 and the American Insensitivity to Food Aid', *Proceedings of the Indian History Congress*, 70, 2009–10, p. 887.
33 Kuromiya, Hiroaki, 'The Soviet Famine of 1932–1933 Revisited', *Europe-Asia Studies*, 60 (4), 2008, pp. 663–75.
34 Messing, Simon D., 'Politics as a Factor in the 1984–1985 Ethiopian Famine', *Africa Today*, 35 (3/4), 1988, p. 100.
35 Our World in Data – Famines: https://ourworldindata.org/famines (impression: 18 September 2020).
36 Mogie, Michael, 'Malthus and Darwin: World Views Apart', *Evolution*, 50 (5), pp. 2086–8.
37 Ehrlich, Paul, *The Population Bomb*, New York, Ballantyne Books, 1968, p. 11.

38 1 News Day, 24 March 2018 https://1newsday.com/world/doomsday-biologist-warns-of-collapse-of-civilization-in-near-future.html (impression: 25 October 2021)
39 Brown, Lester, *Outgrowing the Earth: Food Security and Challenge in an Age of Falling Water Tables and Rising Temperatures*, London, Earthscan, 2005, p. 188.
40 Fuglie, Keith Owen, 'Is Agricultural Productivity Slowing?', *Global Food Security*, 17, 2018, pp. 73–83.
41 Our World In Data: https://ourworldindata.org/employment-in-agriculture (impression: 6 October 2020).
42 Ibid.
43 *Science Magazine*, 21 April 2020: https://www.sciencemag.org/news/2020/04/rice-genetically-engineered-resist-heat-waves-can-also-produce-20-more-grain (impression: 13 October 2020).
44 BBC: https://www.bbc.co.uk/worldservice/specials/119_wag_climate/page10.shtml#:~:text=Well%2C%20it%27s%20cultivated%20on%20six,%2C%20environmental%2C%20political%20and%20cultural (impression: 17 September 2020).
45 Ricepidia: http://ricepedia.org/rice-as-a-crop/rice-productivity (impression: 17 September 2020).
46 FAO: http://www.fao.org/faostat/en/#data/QCL - select regions / world total, elements / production quantity, items / crops primary / rice paddy, years 2000 and 2019(impression: 17 September 2020).
47 Our World in Data: https://ourworldindata.org/hunger-and-undernourishment (impression: 17 September 2020).
48 World Economic Forum 23 July 2020: https://www.weforum.org/agenda/2020/07/global-hunger-rising-food-agriculture-organization-report/ (impression: 1 October 2020).
49 FAO: http://www.fao.org/worldfoodsituation/foodpricesindex/en/ (impression: 1 October 2020).
50 Patel, Raj, *Stuffed and Starved: Markets, Power and the Hidden Battle for the World Food System*, London, Portobello, 2007, p. 1.
51 Al Lahham, Saad et al., 'The Prevalence of Underweight, Overweight and Obesity Among Palestinian School-Aged Children and Associated Risk Factors: A Cross Sectional Study', *BMC Paediatrics*, 19, 2019: https://www.ncbi.nlm.nih.gov/pmc/articles/PMC6902423/ (impression: 17 September 2020).
52 Schwekendiek, Daniel, 'Height and Weight Differences between North and South Korea', *Journal of Biosocial Science*, 41 (1), 2009, pp. 51–5.

53 *Hindustan Times*, 1 August 2017: https://www.hindustantimes.com/india-news/agricultural-output-rose-five-fold-in-60-years-but-farming-sector-is-in-distress/story-cu3zGEbBAb5yB9l2LoJAvN.html (impression: 17 September 2020).
54 Dorling, Danny and Gietel-Basten, Stuart, *Why Demography Matters*, London, Polity, 2018, p. 66.
55 Paltasingh, Kirtti Ranjan and Goyari, Phanindra, 'Impact of Farm Education on Farm Productivity Under Varying Technologies: Case of Paddy Growers in India', *Agricultural and Food Economics*, 6 (1), 2018, pp. 1–19.
56 Vikaspedia: https://vikaspedia.in/agriculture/best-practices/agri-based-enterprises/case-studies-agri-enterprises (impression: 17 September 2020).
57 Farm Radio International, 28 June 2016: https://farmradio.org/mobile-phones-transforming-african-agriculture/ (impression: 1 October 2020).
58 *Financial Times*, 15 October 2018: https://www.ft.com/content/3316885c-b07d-11e8-87e0-d84e0d934341 (impression: 1 October 2020).
59 Blum, Jerome, 'Michael Confino's "Systèmes Agraires et Progrès Agricole"', *The Journal of Modern History*, 43 (3), 1971, pp. 295–8.
60 Business Standard, 2 October 2018: https://www.business-standard.com/article/economy-policy/indian-farm-size-shrank-further-by-6-in-5-years-to-2015-16-census-shows-118100101057_1.html#:~:text=The%20average%20size%20of%20the,census%20released%20on%20Monday%20showed (impression: 1 October 2020).
61 UN Report: https://www.un.org/en/chronicle/article/biotechnology-solution-hunger#:~:text=GM%20crops%20will%20hopefully%20produce,farmers%22%20are%20from%20developing%20countries (impression: 21 September 2020).
62 The Verge, 18 February 2015: https://www.theverge.com/2015/2/18/8056163/bill-gates-gmo-farming-world-hunger-africa-poverty (impression: 21 September 2020).
63 Conway, Gordon, *One Billion Hungry: Can We Feed the World?*, Ithaca, New York, and London, Cornell University Press, 2012, pp. 180–1.
64 Harvard University, 10 August 2015: http://sitn.hms.harvard.edu/flash/2015/will-gmos-hurt-my-body/ (impression: 21 September 2020).
65 See for example Conway, op. cit., pp. 103–24.
66 Bourne, op. cit., pp. 42–52.
67 *Wired*, 13 April 2017: https://www.wired.co.uk/article/underground-hydroponic-farm (impression: 18 September 2020).

68 Growing Underground: http://growing-underground.com/ (impression: 18 September 2020).
69 Food Processing Technology, 16 August 2017: https://www.foodprocessing-technology.com/features/featurehydroponics-the-future-of-farming-5901289/#:~:text=Hydroponics%20has%20the%20potential%20to,places%20where%20space%20is%20scarce (impression: 16 September 2020).
70 Science Focus, 23 May 2019: https://www.sciencefocus.com/future-technology/the-artificial-meat-factory-the-science-of-your-synthetic-supper/ (impression: 18 September 2020).
71 VegNews, 14 July 2019: https://vegnews.com/2019/7/price-of-lab-grown-meat-to-plummet-from-280000-to-10-per-patty-by-2021 (impression: 18 September 2020).
72 Conway, op. cit., p. 194.
73 George, Henry, *Progress and Poverty: An Inquiry into the Cause of Industrial Depression and of Increase of Want with Increase of Wealth – The Remedy*, New York, Sterling Publishing Company, 1879.
74 Churchill, Winston S., *Thoughts and Adventures*, London, Macmillan, 1942, p. 234.

## CONCLUSION: TOMORROW'S PEOPLE

1 See for example Shellengberger, Michael, *Apocalypse Never: Why Environmental Alarmism Hurts Us All*, New York, Harper, 2020 and Lomborg, Bjorn, *False Alarm: How Climate Change Panic Costs us Trillions, Hurts the Poor, and Fails to Fix the Planet*, New York, Basic Books, 2020.
2 Our World in Data: https://ourworldindata.org/natural-disasters#:~:text=Natural%20disasters%20kill%20on%20average,from%200.01%25%20to%200.4%25. (impression: 24 September 2020).
3 Our World in Data: https://ourworldindata.org/war-and-peace (impression: 26 October 2021).
4 Al Jazeera, 24 September 2021, https://www.aljazeera.com/news/2021/9/24/at-least-350000-people-killed-in-syria-war-new-un-count (impression: 26 October 2021).
5 UNHCR: https://data2.unhcr.org/en/situations/syria#_ga=2.91817306.1525884202.1600957949–632148859.1600957949 (impression: 24 September 2020).
6 *New York Times*, 1 January 2018: https://www.nytimes.com/2018/01/01/world/asia/korean-war-history.html (impression: 24 September 2020).

7 *The Economist*, 16 October 2021, p. 21.
8 Barro, Robert J., Ursula, José F. and Weng, Joanna, 'The Corona Virus and the Great Influenza Epidemic: Lessons from the Spanish 'Flu for the Coronavirus's Potential Effects on Mortality and Economic Activity', *American Enterprise Institute*, 1 March 2020, p. 2.
9 *The Times*, 1 October 2020, p. 10.
10 *Guardian*, 7 October 2020: https://www.theguardian.com/world/2020/oct/07/singapore-to-offer-baby-bonus-as-people-put-plans-on-hold-in-covid-crisis?CMP=Share_iOSApp_Other (impression: 8 October 2020); *The Times*, 24 October 2020, p. 13; *The Economist*, 31 October 2020, pp. 61–2.
11 Medical News Today, 24 September 2010: https://www.medicalnewstoday.com/articles/202473#1 (impression: 29 September 2020).
12 Brainerd, Elizabeth and Cutler, David M., 'Autopsy of an Empire: Understanding Mortality in Russia and the Former Soviet Union', *Journal of Economic Perspectives*, 19 (1), 2005, pp. 107–30.
13 For a comprehensive review of the subject see Steele, Andrew, *Ageless: The New Science of Getting Older Without Getting Old*, London, Bloomsbury, 2020.
14 *New Scientist*, 27 September 2016: https://www.newscientist.com/article/2107219-exclusive-worlds-first-baby-born-with-new-3-parent-technique/ (impression: 29 September 2020).
15 Collin, Lindsay, Reisner, Sari L., Tangpricha, Vin and Goodman, Michael, 'Prevalence of Transgender Depends on the "Case" Definition: A Systematic Review', *Journal of Sexual Medicine*, 13 (4), 2016, pp. 613–26.
16 On this subject see, for example, Kurzweil, Ray, *The Singularity is Near: When Humans Transcend Biology*, London, Viking, 2005; Tegmark, Max, *Life 3.0: Being Human in the Age of Artificial Intelligence*, London, Allen Lane, 2017.
17 For a fuller discussion see Mic, 25 May 2020: https://www.mic.com/p/11-brutally-honest-reasons-millenials-dont-want-kids-19629045 (impression: 26 October 2020).
18 OECD, 17 December 2016: https://www.oecd.org/els/family/SF_2_2-Ideal-actual-number-children.pdf (impression: 26 October 2020).
19 Liang, Morita, 'Some Manifestations of Japanese Exclusionism', 13 August 2015: https://journals.sagepub.com/doi/full/10.1177/2158244015600036 (impression: 27 September 2020).

20 For a discussion from the US perspective see for example Borjas, George J., 'Lessons from Immigration Economics', *The Independent Review*, 22 (3), 2018, pp. 329–40.

21 Haaretz, 31 December 2019: https://www.haaretz.com/israel-news/.premium-in-first-for-israel-jewish-fertility-rate-surpasses-that-of-arabs-1.8343039 (impression: 27 September 2020).